Graduate Texts in Mathematics 270

Graduate Texts in Mathematics

Graduate Texts in Mathematics bridge the gap between passive study and creative understanding, offering graduate-level introductions to advanced topics in mathematics. The volumes are carefully written as teaching aids and highlight characteristic features of the theory. Although these books are frequently used as textbooks in graduate courses, they are also suitable for individual study.

More information about this series at http://www.springer.com/series/136

Steven H. Weintraub

Fundamentals of Algebraic Topology

 Springer

Steven H. Weintraub
Department of Mathematics
Lehigh University
Bethlehem, PA, USA

ISSN 0072-5285   ISSN 2197-5612 (electronic)
ISBN 978-1-4939-4885-7   ISBN 978-1-4939-1844-7 (eBook)
DOI 10.1007/978-1-4939-1844-7
Springer New York Heidelberg Dordrecht London

Printed on acid-free paper

Springer is part of Springer Science+Business Media (www.springer.com)

For in much wisdom is much grief, and he that increases knowledge increases sorrow.

Ecclesiastes 1:18

Of making many books there is no end, and much study is a weariness of the flesh.

Ecclesiastes 12:12

Preface

The title of this book, *Fundamentals of Algebraic Topology*, summarizes its aims very well.

In writing this book we have attempted to provide the reader with a guide to the fundamental results of algebraic topology, but we have not attempted to provide an exhaustive treatment.

Our choice of topics is quite standard for an introductory book on algebraic topology, but a description of our approach is in order.

We begin with a short introductory chapter, with basic definitions. We assume the reader is already familiar with basic notions from point-set topology, and take those for granted throughout the book.

We then devote Chap. 2 to the fundamental group, including a careful discussion of covering spaces, van Kampen's theorem, and an application of algebraic topology to obtain purely algebraic results on free groups. In general, algebraic topology involves the use of algebraic methods to obtain topological information, but this is one instance in which the direction is reversed.

We then move on to discuss homology and cohomology. Here we follow the axiomatic approach pioneered by Eilenberg and Steenrod. In Chap. 3 we introduce the famous Eilenberg-Steenrod axioms. Indeed, in this chapter we consider arbitrary generalized homology theories and derive results that hold for all of them. Then, in Chap. 4, we specialize to ordinary homology theory with integer coefficients, and derive results and applications in this situation, still proceeding axiomatically. Of particular note are such important results as the Brouwer fixed point theorem and invariance of domain, which follow from the existence of a homology theory, not from the details of its particular construction. Also of particular note is our introduction of CW complexes and our development of cellular homology, again from the axioms.

Of course, at some point we must show that a homology theory actually exists, and we do that in Chap. 5, where we construct singular homology. We deal with the full panoply here – homology, cohomology, arbitrary coefficients, the Künneth formula, products, and duality.

Manifolds are a particularly important class of topological spaces, and we devote Chap. 6 to their study.

Finally, in Chap. 7 we give a short introduction to homotopy theory.

Arguments in algebraic topology involve a mixture of algebra and topology. But some arguments are purely algebraic, and, indeed, algebraic topology spawned a new branch of algebra, homological algebra, to deal with the algebraic issues it raised. While it is not always possible to completely separate the topology and the algebra, in many instances it is. In those instances, we find it advantageous to do so, as it better reveals the logical structure of the subject. Thus we have included the basic algebraic constructions and results in an appendix, rather than mixing them in with the rest of the text.

In our discussion of Poincaré duality on manifolds, we need some basic facts about bilinear forms, and we summarize them in a second appendix.

Algebraic topology also spawned the language of categories and functors. We have tended to avoid this language in the text, as it is mostly (though not entirely) superfluous for our purposes here. But it is illuminating language, and essential for students who wish to go further, and so we have included a third appendix that introduces it.

There are several points we wish to call the reader's attention to. The first is that we have not felt compelled to give the proofs of all the theorems. To be sure, we give most of them (and leave a few of them as exercises for the reader), but we have omitted some that are particularly long or technical. For example, we have not proved van Kampen's theorem, nor have we proved that singular homology satisfies the excision axiom. The second is that we have not always stated results in the maximum generality. For example, in developing products in homology and cohomology we have restricted the pairs of spaces involved, and the coefficients of the (co)homology groups, to the situations in which they are most often used in practice.

The third is that we have hewed to the axiomatic foundations of the subject. Since its inception over a century ago, algebraic topology has built a vast superstructure on these foundations, a superstructure we hope the student will go on to investigate. But we do not investigate it here. For example, we say very little about techniques for computing homotopy groups.

However, we provide a short bibliography where the reader can find the material we omit, as well as material that is beyond our scope. (In other words, the reader who wants to see this, or wants to study further in algebraic topology, should consult these books.)

Our notation and numbering scheme here are rather standard, and there is little to be said. But we do want to point out that we use $A \subseteq B$ to mean that A is a subset of B and $A \subset B$ to mean that A is a proper subset of B.

Lemmas, propositions, theorems, and corollaries are stated in italics, which clearly delimits them from the text that follows. Similarly, proofs are delimited by the symbol $\square$ at the end. But there is usually nothing to delimit definitions, examples, and remarks, which are stated in roman. We use the symbol $\lozenge$ at the end of these for that purpose.

Bethlehem, PA, USA   Steven H. Weintraub
October 2013

Contents

Chapter 1
The Basics

1.1 Background

As this is a book on topology, all spaces will be topological spaces, and all *maps* (i.e., functions) will be continuous. Basic topological spaces we will consider include:

$\mathbb{R}^n$ n-dimensional Euclidean space

D^n the closed unit disk in $\mathbb{R}^n$

$\mathring{D}^n$ the open unit disk in $\mathbb{R}^n$

S^{n-1} the unit sphere in $\mathbb{R}^n$

S^1 the unit circle in $\mathbb{C}$

I the interval $[0,1]$

$*$ the space consisting of a singe point

$\emptyset$ the empty space

A *pair* (X,A) consists of a space X and a subspace A. We let $f: X \to Y$ denote a map from the space X to the space Y. Similarly, we let $f: (X,A) \to (Y,B)$ denote a map from the pair (X,A) to the pair (Y,B), i.e., a map $f: X \to Y$ whose restriction is $f|A: A \to B$, or more simply, a map $f: X \to Y$ with $f(A) \subseteq B$.

For most purposes we can identify the pair $(X,\emptyset)$ with the space X. For example, with this identification, we could simply have defined $f: (X,A) \to (Y,B)$ and then the definition of $f: X \to Y$ would just have been a special case. (But this logical economy would have been at the expense of clarity.)

A *homeomorphism* $f: X \to Y$ is a continuous map with a continuous inverse $g: Y \to X$, and a homeomorphism $f: (X,A) \to (Y,B)$ is defined similarly.

© Springer International Publishing Switzerland 2014
S.H. Weintraub, *Fundamentals of Algebraic Topology*, Graduate Texts in Mathematics 270, DOI 10.1007/978-1-4939-1844-7_1

In this situation the spaces X and Y, or the pairs (X,A) and (Y,B), are said to be *homeomorphic*, which we write as $X \approx Y$ or $(X,A) \approx (Y,B)$. As basic examples we have:

$$\mathring{D}^n \approx \mathbb{R}^n,$$

$$D^n/S^{n-1} \approx S^n.$$

Here, as usual, for a pair (X,Y) we let X/Y denote the quotient space (of course with the quotient topology). We also have the homeomorphism

$$f : [0,1]/(\{0\} \cup \{1\}) \longrightarrow S^1$$

given by $f(t) = \exp(2\pi it)$, which we will often be (implicitly or explicitly) using.

Given two spaces X and Y, we let $X \times Y$ be their product (of course with the product topology). We define the product of two pairs by

$$(X,A) \times (Y,B) = (X \times Y, X \times B \cup A \times Y).$$

1.2 Homotopy

The basic relation studied in algebraic topology is that of *homotopy*.

Definition 1.2.1. Two maps $f_0 : (X,A) \to (Y,B)$ and $f_1 : (X,A) \to (Y,B)$ are *homotopic*, written $f_0 \sim f_1$, if there is a map

$$F : (X,A) \times I \longrightarrow (Y,B)$$

with $F(x,0) = f_0(x)$ and $F(x,1) = f_1(x)$ for every $x \in X$. ◇

Note that $F : (X,A) \times I \to (Y,B)$ is equivalent to $F : X \to Y$ with $f(a,t) \in B$ for every $a \in A$ and $t \in I$.

It is psychologically very helpful to think of $t \in I$ as "time" and $f_t : (X,A) \to (Y,B)$ by $f_t(x) = f(x,t)$ being "f at time t". With this notion, a homotopy is a deformation through time of f_0 into f_1. It is important to note that while each f_t must be continuous, the condition that F be continuous is stronger than that. For example, the maps $f_0 : \{0,1\} \to \{0,1\}$ given by $f_0(0) = 0$ and $f_0(1) = 1$, and $f_1 : \{0,1\} \to \{0,1\}$ given by $f_1(0) = f_1(1) = 0$ are *not* homotopic, but $F : \{0,1\} \times I \to \{0,1\}$ defined by $f(x,t) = 0$ if $(x,t) \neq (1,1)$ and $F(1,1) = 1$ has each f_t continuous.

Lemma 1.2.2. *Homotopy is an equivalence relation.*

Proof. Reflexive: f is homotopic to f via the homotopy of waiting (i.e., changing nothing) for one unit of time.

Symmetric: If f is homotopic to g, then g is homotopic to f via the homotopy of running the original homotopy backwards in time.

Transitive: If f is homotopic to g and g is homotopic to h, then f is homotopic to h via the homotopy of first doing the original homotopy from f to g twice as fast in the first half of the interval of time, and then doing the original homotopy from g to h twice as fast in the second half of the interval of time. □

We have a closely allied definition.

Definition 1.2.3. Two maps $f_0 : X \to Y$ and $f_1 : X \to Y$ are *homotopic* rel A, where A is a subspace of X, written $f_0 \sim_A f_1$, if there is a map

$$F : X \times I \longrightarrow Y$$

with $F(x,0) = f_0(x)$, $F(x,1) = f_1(x)$ for every $x \in X$, and also $F(a,t) = f_0(a) = f_1(a)$ for every $a \in A$ and every $t \in I$. ◊

In other words, in a homotopy rel A, the points in A never move under the deformation from f_0 to f_1. By exactly the same logic, homotopy rel A is an equivalence relation.

Similar to the relationship of homotopy between maps there is a relationship between spaces.

Definition 1.2.4. Two pairs (X,A) and (Y,B) are *homotopy equivalent* if there are maps $f : (X,A) \to (Y,B)$ and $g : (Y,B) \to (X,A)$ such that the composition $gf : (X,A) \to (X,A)$ is homotopic to the identity map $id : (X,A) \to (X,A)$ and $fg : (Y,B) \to (Y,B)$ is homotopic to the identity map $id : (Y,B) \to (Y,B)$. ◊

As a special case of this we have the following.

Definition 1.2.5. A subspace A of X is a *deformation retract* of X if there is a retraction $g : X \to A$ (i.e., a map $g : X \to A$ with $g(a) = a$ for every $a \in A$) such that g is homotopic to the identity map $id : X \to X$. A subspace A of X is a *strong deformation retract* if there is a retraction $g : X \to A$ such that g is homotopic rel A to the identity map $id : X \to X$. ◊

Lemma 1.2.6. *If A is a deformation retract of X then the inclusion map $i : A \to X$ (defined by $i(a) = a \in X$ for every $a \in A$) is a homotopy equivalence.*

Example 1.2.7. Let us regard $X : \mathbb{R}^n - \{(0,\dots,0)\}$ as the space of nonzero vectors $\{v \in \mathbb{R}^n \mid v \neq 0\}$. Then $A = S^{n-1} = \{v \in \mathbb{R}^n \mid \|v\| = 1\}$ is a subspace of X, and is a strong deformation retract of X. The map

$$F : X \times I \longrightarrow A$$

given by

$$F(v,t) = \|v\|^t (v/\|v\|)$$

gives a homotopy rel A from the retraction $f_0(v) = v/\|v\|$ to the identity map $f_1(v) = v$. $\Diamond$

We have the following common language, which we will use throughout.

Definition 1.2.8. Spaces X and Y that are homotopy equivalent are said to be of the same *homotopy type*. $\Diamond$

Definition 1.2.9. A space X is *contractible* if it has the homotopy type of a point $*$, or, equivalently, if for some, and hence for any, point $x_0 \in X$, x_0 is a deformation retract of X. $\Diamond$

Here is an important construction.

Definition 1.2.10. Let X be a space. The *cone* on X is the quotient space $cX = X \times I/X \times \{1\}$. $\Diamond$

Example 1.2.11. For any $n \geq 1$, cS^{n-1} is homeomorphic to D^n. $\Diamond$

Lemma 1.2.12. *For any space X, cX is contractible.*

Proof. Let $F : cX \times I \to cX$ be defined by

$$F((x,s),t) = (x, \max(s,t)).$$

□

1.3 Exercises

Exercise 1.3.1. (a) Construct a homeomorphism $h : \mathring{D}^n \to \mathbb{R}^n$.
(b) Construct a homeomorphism $h : D^n/S^{n-1} \to S^n$.

Exercise 1.3.2. Let H be the "southern hemisphere" in S^n, $H = \{(x_1,\ldots,x_{n+1}) \in S^n \mid x_{n+1} \leq 0\}$. Let p be the "south pole" $p = (0,0,\ldots,0,-1) \in S^n$. Show that the inclusion $(S^n, p) \to (S^n, H)$ is a homotopy equivalence of pairs.

Exercise 1.3.3. Carefully prove that homotopy is an equivalence relation (Lemma 1.2.2).

Exercise 1.3.4. Prove Lemma 1.2.6.

Exercise 1.3.5. (a) Prove that cS^{n-1} is homeomorphic to D^n (Example 1.2.11).
(b) The *suspension* ΣX of a space X is the quotient space $X \times [-1,1]/\sim$, where the relation $\sim$ identifies $X \times \{1\}$ to a point and $X \times \{-1\}$ to a point.
Prove that ΣS^{n-1} is homeomorphic to S^n.

Chapter 2
The Fundamental Group

One of the basic invariants of the homotopy type of a topological space is its fundamental group. In this chapter we define the fundamental group of a space and see how to calculate it. We also see the intimate relationship between the fundamental group and covering spaces.

Throughout this chapter, all spaces are assumed to be path-connected, unless explicitly stated otherwise.

2.1 Definition and Basic Properties

Definition 2.1.1. Let $x_0 \in X$ be a given point, called the *base point*. The *fundamental group* $\pi_1(X, x_0)$ is the set of homotopy classes of maps

$$f : (S^1, 1) \longrightarrow (X, x_0)$$

with composition given by $h = fg$ where

$$h(e^{i\theta}) = \begin{cases} f(e^{2i\theta}) & 0 \le \theta \le \pi \\ g(e^{2i(\theta-\pi)}) & \pi \le \theta \le 2\pi. \end{cases}$$

Equivalently, $\pi_1(X, x_0)$ is the set of homotopy classes of maps

$$f : (I, \{0\} \cup \{1\}) \longrightarrow (X, x_0)$$

© Springer International Publishing Switzerland 2014
S.H. Weintraub, *Fundamentals of Algebraic Topology*, Graduate Texts
in Mathematics 270, DOI 10.1007/978-1-4939-1844-7_2

with composition given by $h = fg$ where

$$h(t) = \begin{cases} f(2t) & 0 \le t \le \dfrac{1}{2}, \\ g(2t-1) & \dfrac{1}{2} \le t \le 1. \end{cases}$$

◊

We give the equivalence explicitly. We let $S^1 = \{\exp(2\pi it) \mid 0 \le t \le 1\}$ and note that $1 = \exp(2\pi i 0) = \exp(2\pi i 1)$. If $\tilde{f} : (S^1, 1) \to (X, x_0)$ is a map, then we obtain $f : (I, \{0\} \cup \{1\}) \to (X, x_0)$ by $f(t) = \tilde{f}(\exp(2\pi it))$, and if $f : (I, \{0\} \cup \{1\}) \to (X, x_0)$ is a map, then we obtain $\tilde{f} : (S^1, 1) \to (X, x_0)$ by $\tilde{f}(\exp(2\pi it)) = f(t)$.

But in the sequel we will use this identification implicitly and will not distinguish between $\tilde{f}$ and f.

Lemma 2.1.2. *The fundamental group $\pi_1(X, x_0)$ is a group.*

Proof. The identity element of this group is represented by the constant map $i : S^1 \to \{x_0\}$ and the inverse of $f : (S^1, 1) \to (X, x_0)$ is represented by the map $g : (S^1, 1) \to (X, x_0)$ given by $g(e^{i\theta}) = f(e^{-i\theta})$, $0 \le \theta \le 2\pi$ (i.e., the map that runs around the image of S^1 in the opposite direction). □

It is convenient, though admittedly imprecise, to write that the constant path is the identity element of $\pi_1(X, x_0)$, rather than that it represents the identity, and we will often use this language.

As an abstract group, $\pi_1(X, x_0)$ is independent of the choice of the basepoint x_0. More precisely, we have the following result.

Lemma 2.1.3. *Let $x_0, x_1 \in X$. Choose a path φ from x_0 to x_1, i.e., a map $\varphi : I \to X$ with $\varphi(0) = x_0$ and $\varphi(1) = x_1$. Let $\bar{\varphi} : I \to X$ be the map given by $\bar{\varphi}(t) = \varphi(1-t)$, $0 \le t \le 1$. Then there is an isomorphism $\Phi : \pi_1(X, x_1) \to \pi_1(X, x_0)$ given as follows. Let $f : (I, \{0\} \cup \{1\}) \to (X, x_1)$. Then $g = \Phi(f)$ is given by*

$$g(t) = \begin{cases} \varphi(3t) & 0 \le t \le \dfrac{1}{3}, \\ f(3t-1) & \dfrac{1}{3} \le t \le \dfrac{2}{3}, \\ \bar{\varphi}(3t-2) & \dfrac{2}{3} \le t \le 1. \end{cases}$$

Note that this isomorphism depends on the choice of φ. Also, if we choose $x_1 = x_0$, then φ represents an element of $\pi_1(X, x_0)$, and then Φ is the inner automorphism of $\pi_1(X, x_0)$ given by $\Phi(f) = \varphi f \varphi^{-1}$, and conversely every inner automorphism of $\pi_1(X, x_0)$ arises in this way.

A map between spaces induces a map between their fundamental groups as follows.

Definition 2.1.4. Let $f : X \to Y$ with $f(x_0) = y_0$. Then the *induced map* $f_* = \pi_1(X, x_0) \to \pi_1(Y, y_0)$ is defined as follows: Let $g : (S^1, 1) \to (X, x_0)$ represent an element of $\pi_1(X, x_0)$. Then $f_*(g)$ is the element of $\pi_1(Y, y_0)$ represented by the composition $f \circ g : (S^1, 1) \to (Y, y_0)$. $\diamondsuit$

It must be checked that f_* is well-defined, i.e., independent of the choice of representative g, but this is routine.

Clearly if f is a homeomorphism, then f_* is an isomorphism. But we have a more general result.

Theorem 2.1.5. *Let $f : X \to Y$ with $f(x_0) = y_0$. If f is a homotopy equivalence, then $f_* : \pi_1(X, x_0) \to \pi_1(Y, y_0)$ is an isomorphism.*

Proof. Suppose that $g : Y \to X$ with $g(y_0) = x_0$, that $gf : X \to X$ is homotopic to the identity rel x_0, and that $fg : Y \to Y$ is homotopic to the identity rel y_0. Then the theorem is very easy to prove. But we are not making that strong an assumption, and so the proof is trickier, but we still leave it to the reader. $\square$

Corollary 2.1.6. *Let X be a contractible space. Then $\pi_1(X, x_0)$ is the trivial group.*

Proof. This is clearly true if X is the space consisting of the point x_0 alone, as then every $f : (S^1, 1) \to (X, x_0)$ is the constant map to the point x_0. Then it is also true for X contractible by Theorem 2.1.5. $\square$

Theorem 2.1.7. *Let $x_0 \in X$ and $y_0 \in Y$. Then $\pi_1(X \times Y, (x_0, y_0))$ is isomorphic to the product $\pi_1(X, x_0) \times \pi_1(Y, y_0)$.*

Proof. Let p be the projection $p : X \times Y \to X$ and let q be the projection $q : X \times Y \to Y$. It is easy to check that $p_* \times q_* : \pi_1(X \times Y, (x_0, y_0)) \to \pi_1(X, x_0) \times \pi_1(Y, y_0)$ is a homomorphism. To show that it is onto, let $f : (S^1, 1) \to (X, x_0)$ represent an arbitrary element α of $\pi_1(X, x_0)$, and let $g : (S^1, 1) \to (Y, y_0)$ represent an arbitrary element β of $\pi_1(Y, y_0)$. Let $h = f \times g : (S^1, 1) \to (X \times Y)$, i.e., $h(t) = (f(t), g(t))$. Then h represents an element γ of $\pi_1(X \times Y, (x_0, y_0))$ and $p_* \times q_*(\gamma) = \alpha \times \beta$. To show that it is one-to-one, let $h : (S^1, 1) \to (X \times Y, (x_0, y_0))$ represent $\gamma \in \pi_1(X \times Y, (x_0, y_0))$ and suppose $p_* \times q_*(\gamma)$ is the identity element of $\pi_1(X, x_0) \times \pi_1(Y, y_0)$, i.e., that $f = p(h) : (S^1, 1) \to (X, x_0)$ and $g = q(h) : (S^1, 1) \to (Y, y_0)$ are both null-homotopic. Let $F : (S^1, 1) \times I \to (X, x_0)$ and $G : (S^1, 1) \times I \to (Y, y_0)$ be homotopies between f and the constant map to x_0 and between g and the constant map to y_0, respectively. Then $H = F \times G : (S^1, 1) \times I \to X \times Y$ is a homotopy between h and the constant map to (x_0, y_0), so γ is the identity element of $\pi_1(X \times Y, (x_0, y_0))$. $\square$

Definition 2.1.8. The path-connected space X is *simply connected* if for some (and hence any) point $x_0 \in X$ the fundamental group $\pi_1(X, x_0)$ is trivial. $\diamondsuit$

We have yet to show that there are spaces X that are not simply connected. We do that, and more, in the next two sections, where we develop techniques for calculating fundamental groups.

2.2   Covering Spaces

Definition 2.2.1. Let X and Y be arbitrary spaces, and let $p : Y \to X$ be a map. Then Y is a *covering space* of X, and p is a *covering projection*, if every $x \in X$ has an open neighborhood U with $p^{-1}(U) = \{V_i\}$ a union of open sets in Y, and with $p|V_i : V_i \to U$ a homeomorphism for each i. Such a set U is said to be *evenly covered* by p. $\Diamond$

Lemma 2.2.2. *Let $p : Y \to X$ be a covering projection. Then*

(i) *For every $x \in X$, $\{p^{-1}(x)\}$ is a discrete subset of Y.*
(ii) *p is a local homeomorphism.*
(iii) *The topology on X is the quotient topology it inherts from Y via the map p.*

We have defined covering spaces in complete generality. But in order to obtain a relationship between covering spaces and the fundamental group, we shall have to assume that both X and Y are path connected. Indeed, to get the best relationship we shall have to restrict our attention even further. But for now we continue in general.

Example 2.2.3.   (i) Let D be any discrete space. Then $X \times D$ is a covering space of X, with the covering projection p being a projection onto the first factor.

(iia) Let $p : \mathbb{R} \to S^1$ be defined by $p(t) = e^{2\pi i t}$. Then p is a covering projection.

(iib) Let n be a positive integer and let $p : S^1 \to S^1$ be defined by $p(z) = z^n$. Then p is a covering projection.

(iii) Let G be any topological group and let H be any discrete subgroup of G. Then the projection $p : G \to H \backslash G$ (the space of left cosets, with the quotient topology) is a covering projection.

(iv) Let Y be any Hausdorff topological space and let G be any finite group that acts freely on Y, i.e., with the property that if $g(y) = y$ for any $g \in G$ and any $y \in Y$, then g is the identity element of G. Let $G \backslash Y$ be the quotient space under this action, i.e., $y_1, y_2 \in Y$ are identified in $G \backslash Y$ if there is an element g of G with $g(y_1) = y_2$, with the quotient topology. Note we are assuming here that G acts on Y on the left. Then $p : Y \to G \backslash Y$ is a covering projection. More generally, let Y be any topological space and let G be any group acting properly discontinuously on Y, i.e., with the property that every $y \in Y$ has a neighborhood U such that if $g(U) \cap U \neq \emptyset$, then g is the identity element of G. (Note that such an action must be free.) Then $p : Y \to G \backslash Y$ is a covering projection.

Note that (ii) is a special case of (iii), which is in turn a special case of (iv). $\Diamond$

A covering projection $p : Y \to X$ has two important properties.

Theorem 2.2.4 (Unique path lifting).   *Let $p : Y \to X$ be a covering projection. Let $x_0 \in X$ be arbitrary and let $y_0 \in Y$ be any point with $p(y_0) = x_0$. Let $f : I \to X$ be an arbitrary map with $f(0) = x_0$. Then f has a unique lifting $\tilde{f} : I \to Y$ with $\tilde{f}(0) = y_0$, i.e., there is a unique $\tilde{f} : I \to Y$ with $\tilde{f}(0) = y_0$ making the following diagram commute:*

Theorem 2.2.5 (Homotopy lifting property). *Let $p : Y \to X$ be a covering projection. Let E be an arbitrary space and let $F : E \times I \to X$ be an arbitrary map. Suppose there is a map $\tilde{f} : E \times \{0\} \to Y$ such that $p\tilde{f}(e,0) = F(e,0)$ for every $e \in E$. Then $\tilde{f}$ extends to a map $\tilde{F} : E \times I \to Y$ making the following diagram commute:*

The homotopy lifting property is sometimes also called the covering homotopy property.

As a consequence of these two theorems, we can now compute some fundamental groups.

Theorem 2.2.6. *Let Y be a path connected and simply connected space and let the group G act properly discontinuously on Y on the left. Let X be the quotient space $X = G \backslash Y$. Then for any $x_0 \in X$, $\pi_1(X, x_0)$ is isomorphic to G.*

Proof. Let $p : Y \to X$ be the quotient map. As we have observed, p is a covering projection. Choose $y_0 \in Y$ with $p(y_0) = x_0$. Note that $g \mapsto y_g = g(y_0)$ gives a 1-1 correspondence between the elements g of G and $F = \{y \in Y \mid p(y) = x_0\}$, and under this correspondence $y_e = y_0$, where e is the identity element of G. For each $g \in G$, define $\tilde{f}_g : I \to Y$ with $\tilde{f}_g(0) = y_0$ and $\tilde{f}_g(1) = y_g$. Note that such a map, $\tilde{f}_g$ always exists as we are assuming Y is path connected. Then $f_g = p(\tilde{f}_g)$ is a map $f_g : I \to X$ with $f_g(0) = f_g(1) = x_0$, so represents an element $[f_g]$ of $\pi_1(X, x_0)$. We claim the map $\varphi : G \to \pi_1(X, x_0)$ by $\varphi(g) = [f_g]$ is an isomorphism.

(i) φ is well-defined. Suppose we have another map $\tilde{f}'_g : I \to Y$ with $\tilde{f}'_g(0) = y_0$ and $\tilde{f}'_g(1) = y_g$. Since Y is simply connected, $\tilde{f}_g$ and $\tilde{f}'_g$ are homotopic rel $\{0,1\}$, so taking the image of this homotopy under p gives a homotopy between f_g and f'_g rel $\{0,1\}$, so $[f_g] = [f'_g] \in \pi_1(X, x_0)$.

(ii) φ is onto. Let $\alpha \in \pi_1(X, x_0)$ be arbitrary and let $f : (I, \{0,1\}) \to (X, x_0)$ represent α. Then by Theorem 2.2.4 f lifts to $\tilde{f} : I \to Y$ with $\tilde{f}(0) = y_0$ and $p(\tilde{f}(1)) = x_0$, i.e., $\tilde{f}(1) = y_g$ for some element g of G, and then by (i) we may take $\tilde{f}_g = \tilde{f}$, so $\varphi(\tilde{f}) = g$.

(iii) φ is one-to-one. Suppose that for some $g \in G$, $f_g = \pi(\tilde{f}_g)$ represents the trivial element of $\pi_1(X, x_0)$, i.e., it is null-homotopic rel $\{0,1\}$. Then there is a map $F : I \times I \to X$ with $F(s,0) = f_g(s) = p\tilde{f}_g(s)$ for every $s \in I$, $F(0,t) = F(1,t) = x_0$ for every $t \in I$ and $F(s,1) = x_0$ for every $s \in I$. By Theorem 2.2.5, F lifts to

$\tilde{F}: I \times I \to Y$ with $\tilde{F}(s,0) = \tilde{f}_g(s)$ for every $s \in I$, and $p\tilde{F}(0,1) = p\tilde{F}(1,t) = x_0$. Now $\tilde{h}_0 : I \to Y$ by $\tilde{h}_0(t) = \tilde{F}(0,t)$ is a path in Y, and $\tilde{h}_0 = \tilde{F}(0,0) = \tilde{f}_g(0) = y_0$. Also, $p\tilde{h}_0(t) = x_0$ for every t, i.e., $\tilde{h}_0(t) \in p^{-1}(x_0)$ for every $t \in I$. But $p^{-1}(x_0)$ is a discrete space, so we must have $\tilde{h}_0(t) = \tilde{h}_0(0)$ for every t, and in particular $\tilde{h}_0(1) = \tilde{h}_0(0) = y_0$. By exactly the same logic, if $\tilde{h}_1 : I \to Y$ by $\tilde{h}_1(t) = \tilde{F}(1,t)$, we must have $\tilde{h}_1(1) = \tilde{h}_1(0)$. Now $\tilde{h}_1(0) = \tilde{F}(1,0) = \tilde{f}(1) = y_g$. On the other hand, $F(s,1)$ is a constant path starting at x_0, so by Theorem 2.2.4 lifts to a unique path starting at $\tilde{F}(0,1) = y_0$. Obviously, the constant path at y_0 is such a lifting, so must be the only lifting, i.e., $\tilde{F}(s,1) = y_0$ for every s. In particular $\tilde{F}(1,1) = \tilde{h}_1(1) = y_0$. But then $y_0 = y_g = g(y_0)$, and so g is the identity element of G (as the action of G on Y is free).

(iv) φ is a homomorphism. Let $\tilde{f}_{g_1} : I \to Y$ be a path from $\tilde{f}_{g_1}(0) = y_e$ to $\tilde{f}_{g_1}(1) = y_{g_1}$, and let $\tilde{f}_{g_2} : I \to Y$ be a path from $\tilde{f}_{g_2}(0) = y_e$ to $\tilde{f}_{g_2}(1) = y_{g_2}$. Let $p(\tilde{f}_{g_1}) = f_{g_1}$ and $p(\tilde{f}_{g_2}) = f_{g_2}$, and let α_1 and α_2 be the elements of the fundamental group $\pi_1(X, x_0)$ represented by f_{g_1} and f_{g_2}, respectively. Let $\beta = \alpha_1 \alpha_2$. Then β is represented by a loop $f : (I, \{0,1\}) \to (X, x_0)$, which lifts to a path $\tilde{f} : (I, \{0\}) \to (Y, y_0)$, and $\tilde{f}(1) = y_g$ for some $g \in G$. We must show that $g = g_1 g_2$. But here is a path covering f: Let

$$\tilde{f}(t) = \begin{cases} \tilde{f}_{g_1}(2t) & 0 \leq t \leq \dfrac{1}{2}, \\[2mm] g_1(\tilde{f}_{g_2}(2t-1)) & \dfrac{1}{2} \leq t \leq 1. \end{cases}$$

Note this is indeed a path as for $t = 1/2$ we have

$$g_1(\tilde{f}_{g_2}(2t-1)) = g_1(\tilde{f}_{g_2}(0)) = g_1(y_0) = y_{g_1} = \tilde{f}_{g_1}(1).$$

But then

$$y_g = \tilde{f}(1) = g_1(\tilde{f}_{g_2}(1)) = g_1(y_{g_2}) = g_1(g_2(y)) = (g_1 g_2)(y) = y_{g_1 g_2}$$

so $g = g_1 g_2$ as claimed. $\square$

Example 2.2.7. Let $G = \mathbb{Z}$ act on $Y = \mathbb{R}$ by $n(r) = r + n$, $n \in G$ and $r \in \mathbb{R}$. Let $X = G \backslash Y$. Then X is homeomorphic to S^1, so we see that $\pi_1(S^1, 1) = \mathbb{Z}$. Note that we may identify the covering projection $p : Y \to X$ with the covering projection in Example 2.2.3(iia).

It is worth being completely explicit here. Let $\pi : \mathbb{R} \to S^1$ by $\pi(t) = \exp(2\pi i t)$. Let d be an integer and let $\tilde{f}_d : I \to \mathbb{R}$ by $\tilde{f}_d(t) = dt$. Then we have a commutative diagram so $f_d = \pi \tilde{f}_d : (I, \partial I) \to (S^1, 1)$ represents $d \in \pi_1(S^1, 1) \approx \mathbb{Z}$. Note that under the identification of $\pi_1(X, x_0)$ with the set of homotopy classes of maps $f : (S^1, 1) \to (X, x_0)$, the map f_d is identified with the map $z \mapsto z^d$. (In particular, the identity map from S^1 to itself represents a generator $1 \in \pi_1(S^1, 1)$.) ◇

We will see additional examples in the next section.

Here is an important property of covering spaces, which is a generalization of Theorem 2.2.4.

Theorem 2.2.8. *Let $p : Y \to X$ be a covering projection and let $y_0 \in Y$ and $x_0 \in X$ be points with $p(y_0) = x_0$. Let E be an arbitrary connected and locally path connected space and let e_0 be a point in E.*

Let $f : (E, e_0) \to (X, x_0)$ be a map. Then there is a lifting $\tilde{f} : (E, e_0) \to (Y, y_0)$ if and only if

$$f_* = (\pi_1(E, e_0)) \subseteq p_*(\pi_1(Y, y_0))$$

and, if so, $\tilde{f}$ is unique.

(Since $f = p\tilde{f}$, the condition in the theorem is obviously necessary. The point of the theorem is that it is sufficient.)

Definition 2.2.9. Covering projections $p_1 : (Y_1, y_0^1) \to (X, x_0)$ and $p_2 : (Y_2, y_0^2) \to (X, x_0)$ are *equivalent* if there is a homeomorphism $f : (Y_1, y_0^1) \to (Y_2, y_0^2)$ making the following diagram commute:

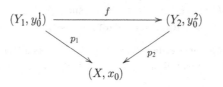

◇

Corollary 2.2.10. *Let Y_1 and Y_2 be path connected and let $p_1 : (Y_1, y_0^1) \to (X, x_0)$ and $p_2 : (Y_2, y_0^2) \to (X, x_0)$ be covering projections. Then (Y_1, y_0^1) and (Y_2, y_0^2) are equivalent if and only if*

$$(p_1)_*(\pi_1(Y_1, y_0^1)) = (p_2)_*(\pi_1(Y_2, y_0^2)).$$

Definition 2.2.11. A covering projection $\tilde{p} : \tilde{X} \to X$ is a *universal cover* if for any covering projection $p : Y \to X$, $\tilde{p}$ lifts to a map $q : \tilde{X} \to Y$. ◇

It is easy to check that in this case q itself must be a covering projection, so we have a tower of covering projections

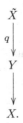

Corollary 2.2.12. *Suppose that X has a simply connected covering space $\tilde{X}$. Then the covering projection $\tilde{p} : \tilde{X} \to X$ is a universal cover.*

Definition 2.2.13. Let $p : Y \to X$ be a covering space with Y path connected. Then the group of *covering translations* (or *deck transformations*) G_p is the group of homeomorphisms $f : Y \to Y$ making the following diagram commute:

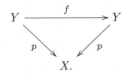

Lemma 2.2.14. *Let y_0 be any point of Y. Then $f \in G_p$ is determined by $f(y_0)$.*

Corollary 2.2.15. *G_p acts properly discontinuously on Y.*

Definition 2.2.16. Let $p : Y \to X$ be a covering projection with Y path connected. Let $x_0 \in X$. For $f : (I, \{0, 1\}) \to (X, x_0)$ and $y \in Y$ with $p(y) = x_0$, let $\tilde{f}_y : (I, 0) \to (Y, y)$ be the unique lift of f given by Theorem 2.2.4.

Then Y is a *regular* cover of X if for each such f, either $f_y(1) = y$ for every y with $p(y) = x_0$ or $f_y(1) \neq y$ for every y with $p(y) = x_0$.

Informally, Y is regular if either every lift of a loop is a loop, or no lift of a loop is a loop. (It is easy to check that the condition of being regular is independent of the choice of x_0.)

Although some results are true more generally, henceforth we shall assume that for a covering projection $p : Y \to X$:

Hypotheses 2.2.17.   (i) *X is connected.*
(ii) *X is locally path connected, i.e., for any $x \in X$ and any neighborhood U of X, there is an open neighborhood $V \subset U$ of x such that V is path connected.*
(iii) *X is semilocally simply connected, i.e., for any $x \in X$ and any neighborhood U of X, there is an open neighborhood $V \subset U$ of x such that $i_* : \pi_1(V, x) \to \pi_1(U, x)$ is the trivial map, where $i : V \to U$ is the inclusion.*
(iv) *Y is connected.*

Note that (i) and (ii) imply that X is path connected. Note that since p is a local homeomorphism, properties (ii), (iii), and (iv) hold for Y as well, and in particular Y is also path connected.

For example, we see that these hypotheses are satisfied for the covering projections in Example 2.2.3(ii). (However, X may be connected without Y being connected, as in Example 2.2.3(i).)

Definition 2.2.18. The degree of the cover is the cardinality of $p^{-1}(x_0)$. ◊

Theorem 2.2.19. *Under Hypotheses 2.2.17:*

(i) *To each subgroup H of $\pi_1(X,x_0)$ there corresponds a covering projection $p : Y \to X$ and a point $y_0 \in Y$ with $p(y_0) = x_0$ such that*

$$p_*(\pi_1(Y,y_0)) = H \subseteq \pi_1(X,x_0)$$

and (Y,y_0) is unique up to equivalence.
(ii) *The points in $p^{-1}(x_0)$ are in 1-1 correspondence with the right cosets of H in $\pi_1(X,x_0)$. Thus the degree of the cover is the index of H in $\pi_1(X,x_0)$.*
(iii) *H is normal in $\pi_1(X,x_0)$ if and only if Y is a regular cover. In this case the group of covering translations is isomorphic to the quotient group $\pi_1(X,x_0)/H$.*

Remark 2.2.20. By Corollary 2.2.10, this is a 1-1 correspondence. ◊

Corollary 2.2.21. *Under Hypotheses 2.2.17:*
Every X has a simply-connected cover $p : \tilde{X} \to X$, unique up to equivalence. $\tilde{X}$ is the universal cover of X, and X is the quotient of $\tilde{X}$ by the group of covering translations. Also, if Y is any cover of X, then $\tilde{X}$ is a cover of Y.

Proof. This is a direct consequence of Theorem 2.2.19, and our earlier results, taking H to be the trivial subgroup of $\pi_1(X,x_0)$. □

Remark 2.2.22. This shows that, in the situation where Hypotheses 2.2.17 hold, the covering projection $p : \tilde{X} \to X$ from the universal cover $\tilde{X}$ to X is exactly the quotient map under the action of the group G_p of covering translations, isomorphic to $\pi_1(X,x_0)$, considered in Theorem 2.2.6.

The only difference is that we have reversed our point of view: In Theorem 2.2.6 we assumed G_p was known, and used it to find $\pi_1(X,x_0)$, while in Theorem 2.2.19 we assumed $\pi_1(X,x_0)$ was known, and used it to find G_p. ◊

2.3 van Kampen's Theorem and Applications

van Kampen's theorem allows us, under suitable circumstances, to compute the fundamental group of a space from the fundamental groups of subspaces.

Theorem 2.3.1. *Let $X = X_1 \cup X_2$ and suppose that X_1, X_2, and $A = X_1 \cap X_2$ are all open, path connected subsets of X. Let $x_0 \in A$. Then $\pi_1(X, x_0)$ is the free product with amalgamation*

$$\pi_1(X, x_0) = \pi_1(X_1, x_0) *_{\pi_1(A, x_0)} \pi_1(X_2, x_0).$$

In other words, if $i_1 : A \to X_1$ and $i_2 : A \to X_2$ are the inclusions, then $\pi_1(X, x_0)$ is the free product $\pi_1(X_1, x_0) * \pi_1(X_2, x_0)$ modulo the relations $(i_1)_*(\alpha) = (i_2)_*(\alpha)$ for every $\alpha \in \pi_1(A, x_0)$.

As important special cases we have:

Corollary 2.3.2. *Under the hypotheses of van Kampen's theorem:*

(i) *If X_1 and X_2 are simply connected, then X is simply connected.*
(ii) *If A is simply connected, then $\pi_1(X, x_0) = \pi_1(X_1, x_0) * \pi_1(X_2, x_0)$.*
(iii) *If X_2 is simply connected, then $\pi_1(X, x_0) = \pi_1(X_1, x_0) / \langle \pi_1(A, x_0) \rangle$ where $\langle \pi_1(A, x_0) \rangle$ denotes the subgroup normally generated by $\pi_1(A, x_0)$.*

Corollary 2.3.3. *For $n > 1$, the n-sphere S^n is simply connected.*

Proof. We regard S^n as the unit sphere in $\mathbb{R}^{n+1}$. Let $X_1 = S^n - \{(0, 0, \ldots, 0, 1)\}$ and $X_2 = S^n - \{(0, 0, \ldots, 0, -1)\}$. Then X_1 and X_2 are both homeomorphic to $\mathring{D}^n$, so are path connected and simply connected, and $X_1 \cap X_2$ is path connected, as $n > 1$, so by Corollary 2.3.2(i) S^n is simply connected. $\qquad\square$

Example 2.3.4. (i) Regard S^n as the unit sphere in $\mathbb{R}^{n+1}$ and let $\mathbb{Z}_2$ act on S^n, where the nontrivial element g of $\mathbb{Z}_2$ acts via the antipodal map, $g(z_1, \ldots, z_{n+1}) = (-z_1, \ldots, -z_{n+1})$. The quotient $\mathbb{R}P^n = S_n / \mathbb{Z}_2$ is *real projective n-space*. Note that $p : S^0 \to \mathbb{R}P^0$ is the map from the space of two points to the space of one point, and $p : S^1 \to \mathbb{R}P^1$ may be identified with the cover in Example 2.2.3(iib) for $n = 2$. But for $n > 1$, by Corollary 2.3.3 and Theorem 2.2.6 we see that $\pi_1(\mathbb{R}P^n, x_0) = \mathbb{Z}_2$.

(ii) For $n = 2m - 1$ odd, regard S^n as the unit sphere in $\mathbb{C}^m$. Fix a positive integer k and integers $j_1, \ldots, j_m$ relatively prime to k. Let the group $\mathbb{Z}_k$ act on S^n where a fixed generator g acts by $g(z_1, \ldots, z_m) = (\exp(2\pi i j_1 / k) z_1, \ldots, \exp(2\pi i j_m / k) z_m)$. The quotient $L = L^{2m-1}(k; j_1, \ldots, j_m)$ is a *lens space*. For $m = 1$ the projection $p : S^{2m-1} \to L$ may be identified with the cover in Example 2.2.3(iib) with (in the notation there) $n = k$. But for $m > 1$, by Corollary 2.3.3 and Theorem 2.2.6 we see that $\pi_1(L, x_0) = \mathbb{Z}_k$. $\qquad\diamond$

Example 2.3.5. Regard S^1 as the unit circle in $\mathbb{C}$. Let n be a positive integer. For $k = 1, \ldots, n$ let $(S^1)_k$ be a copy of S^1. The *n-leafed rose* is the space R_n obtained from the disjoint union of $(S^1)_1, \ldots, (S^1)_n$ by identifying the point 1 in each copy of S^1. $\qquad\diamond$

Let $r_0 \in R_n$ be the common identification point. Let $(S^1)_k$ be coordinated by $(z)_k$, and let $i_k : (S^1)_k \to R_n$ be the inclusion.

Corollary 2.3.6. *The fundamental group $\pi_1(R_n, r_0)$ is the free group on the n elements $\alpha_k = (i_k)_*(g_k)$, where g_k is a generator of $\pi_1((S^1)_k, (1)_k)$, for $k = 1, \ldots, n$.*

Proof. We proceed by induction on n.

For $n = 1$ this is Example 2.2.7.

Now suppose that $n \geq 1$ and that the theorem is true for n. Write $R_{n+1} = X_1 \cup X_2$ where:

$$X_1 = \bigcup_{k=1}^{n} (S^1)_k \cup \{(z)_{n+1} \in (S^1)_{n+1} \mid \mathrm{Re}\,(z) > 0\},$$

$$X_2 = \bigcup_{k=1}^{n} \{(z)_k \in (S^1)_k \mid \mathrm{Re}\,(z) > 0\} \cup (S^1)_{n+1},$$

Then $X_1 \cap X_2$ is contractible. (It has the point 1 as a strong deformation retract.) Also, X_1 has R_n as a strong deformation retract, and X_2 has $(S^1)_{n+1}$ as a strong deformation retract. By the induction hypothesis $\pi_1(R_n, r_0)$ is the free group with generators $\alpha_1, \ldots, \alpha_n$, and by the $n = 1$ case $\pi_1((S^1)_{n+1}, 1)$ is the free group on α_{n+1}, so, by Corollary 2.3.2(ii), $\pi_1(R_{n+1}, r_0)$ is the free group with generators $\alpha_1, \ldots, \alpha_{n+1}$, so the theorem is true for $n + 1$.

Thus by induction we are done.

Here is a picture for the case $n = 3$:

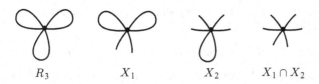

$$R_3 \qquad\qquad X_1 \qquad\qquad X_2 \qquad X_1 \cap X_2$$

$\square$

2.4 Applications to Free Groups

We now show how to use the topological methods we have developed so far to easily derive purely algebraic results about subgroups of free groups.

Definition 2.4.1. A *1-complex* is an identification space $C = (V, E)/\sim$ where $V = \{v_i\}$ is a collection of points, the *vertices* of C, and $E = \{I_j\}$ is a collection of

16   2   The Fundamental Group

intervals, $I = [0,1]$, the *edges* of C, with $0 \in E_j$ identified with some v_i and $1 \in E_j$ identified with some v_i, with C having the quotient topology. C is a *finite complex* if it has finitely many vertices and edges.   ◊

Example 2.4.2.   (i) The n-leafed rose R_n is a 1-complex with 1 vertex and n edges.
(ii) An n-gon, i.e., a polygon with n sides, $n \geq 3$, may be naturally regarded as a
    1-complex with n edges and n vertices.   ◊

A subcomplex of C is a subset of C that is a complex, with vertices a subset of V and edges a subset of E. A tree is a contractible 1-complex. A maximal tree in C is a subcomplex of C that is a tree, and is maximal (under inclusion) among such subcomplexes of C. It is easy to see that a maximal tree always exists, though it may not be unique. Also, if C is connected, a maximal tree must contain all vertices.

Example 2.4.3.   (i) An n-leafed rose has its single vertex as a maximal tree.
(ii) The subcomplex obtained by deleting any edge is a maximal tree in an
    n-gon.   ◊

Theorem 2.4.4. *Let C be a connected 1-complex, v a vertex of C, and T a maximal tree in C. Then $\pi_1(C,v)$ is a free group with generators in 1-1 correspondence with the edges of C not in T. A generator is obtained from such an edge E as follows: Let E have vertices v_{i_0} and v_{i_1}, respectively. Then the generator α_E is represented by a loop in E that consists of a path in T from v to v_{i_0}, then E from v_{i_0} to v_{i_1}, then a path in T from v_{i_1} back to v.*

Note that this theorem generalizes Corollary 2.3.6 in the case of a finite 1-complex, and can be proved in a very similar fashion. It remains true for infinite 1-complexes as well.

Corollary 2.4.5. *Let H be a subgroup of a free group G. Then H is a free group.*

Proof. Consider a rose R whose edges are in 1-1 correspondence with the generators of G. By Theorem 2.2.19, there is a cover $\tilde{R}$ with $\pi_1(\tilde{R},\tilde{v}) = H$. But it is easy to see that, since a covering projection $p : \tilde{R} \to R$ is a local homeomorphism, $\tilde{R}$ is a 1-complex as well. But then $\pi_1(\tilde{R},\tilde{v})$ is free by Theorem 2.4.4.   □

Corollary 2.4.6. *Let G be a free group on k elements and let H be a subgroup of G of index n. Then H is a free group on $(k-1)n+1$ elements.*

Proof. We may consider G to be the fundamental group $\pi_1(R,v)$, where R is a k-leafed rose. Then H is the fundamental group of an n-fold cover $\pi_1(\tilde{R},\tilde{v})$. Now R has 1 vertex and k edges, so $\tilde{R}$ has n vertices and kn edges. Then a maximal tree T in $\tilde{R}$ has $n-1$ edges so $\tilde{R}$ has $kn - (n-1) = (k-1)n+1$ edges not in T and the corollary follows immediately from Theorem 2.4.4.   □

a b

Example 2.4.7. We consider the free group $\langle a, b \rangle$ on two generators a and b, which we regard as the fundamental group of R_2:

(i) We find all subgroups of index 2. These are fundamental groups of 2-fold covers. We give "pictures" of these covers, with each curve or loop on the left (resp. right), traversed in the direction of the arrows, projecting to a (resp. b). Note that we have two choices of basepoint, giving us two possible subgroups, which will be conjugate. But every 2-fold cover is regular (or every subgroup of index 2 is normal), so the choice of basepoint does not matter. We give the covers and the corresponding subgroups. Note each is a free group on three generators.

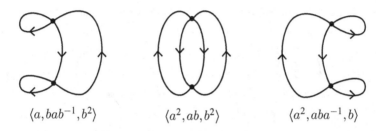

$$\langle a, bab^{-1}, b^2 \rangle \qquad\qquad \langle a^2, ab, b^2 \rangle \qquad\qquad \langle a^2, aba^{-1}, b \rangle$$

(ii) Similarly we find all subgroups of index 3 by finding fundamental groups of 3-fold covers. We give similar pictures. In case the cover is regular, the corresponding subgroup is normal, and we have listed it once, and noted that fact. In case it is not regular, we have listed the different conjugacy classes corresponding to the choices of different basepoints, from top to bottom. Note each is a free group on four generators.

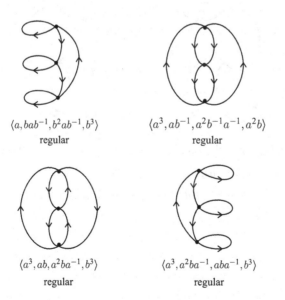

$$\langle a, bab^{-1}, b^2ab^{-1}, b^3 \rangle \qquad\qquad \langle a^3, ab^{-1}, a^2b^{-1}a^{-1}, a^2b \rangle$$
$$\text{regular} \qquad\qquad\qquad\qquad\qquad \text{regular}$$

$$\langle a^3, ab, a^2ba^{-1}, b^3 \rangle \qquad\qquad \langle a^3, a^2ba^{-1}, aba^{-1}, b^3 \rangle$$
$$\text{regular} \qquad\qquad\qquad\qquad\qquad \text{regular}$$

◇

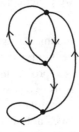

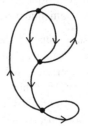

$\langle b^2 ab^{-2}, a^2, ab^2, b^3 \rangle$

$\langle bab^{-1}, a^2, ab, b^3 \rangle$

$\langle a, b^{-1}a^2b, b^{-1}ab^2, b^3 \rangle$

$\langle a^2 ba^{-2}, b^2, ba^2, a^3 \rangle$

$\langle aba^{-1}, b^2, ba, a^3 \rangle$

$\langle b, a^{-1}b^2a, a^{-1}ba^2, a^3 \rangle$

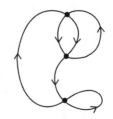

$\langle a^2, abab^{-1}a^{-1}, ab^{-1}, b^3 \rangle$

$\langle a^2, bab^{-1}, ab^{-1}, b^3 \rangle$

$\langle b^{-1}a^2b, a, b^{-1}a^{-1}b^2, b^3 \rangle$

$\langle b^2, baba^{-1}b^{-1}, ba^{-1}, a^3 \rangle$

$\langle b^2, aba^{-1}, ba^{-1}, a^3 \rangle$

$\langle a^{-1}b^2a, b, a^{-1}b^{-1}a^2, a^3 \rangle$

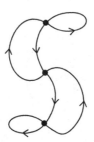

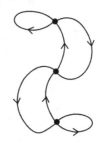

$\langle a^2, abab^{-1}a^{-1}, b, ab^2a^{-1} \rangle$

$\langle a^2, bab^{-1}, a^{-1}ba, b^2 \rangle$

$\langle b^{-1}a^2b, a, b^{-1}a^{-1}bab, b^2 \rangle$

$\langle b^2, baba^{-1}b^{-1}, a, ba^2b^{-1} \rangle$

$\langle b^2, aba^{-1}, b^{-1}ab, a^2 \rangle$

$\langle a^{-1}b^2a, a^{-1}b^{-1}aba, a^2 \rangle$

2.5 Free Homotopy Classes

The fundamental group $\pi_1(X, x_0)$ is, as a set, the set of homotopy classes of maps $(S^1, 1) \to (X, x_0)$. These maps are based at x_0. We now wish to study homotopy classes of maps $S^1 \to X$, known as *free* homotopy classes.

Theorem 2.5.1. *Let X be a path connected space. Then there is a 1-1 correspondence between conjugacy classes of elements of $\pi_1(X, x_0)$ and $\pi(X) = \{$homotopy classes of maps $S^1 \to X\}$.*

Proof. Let $\Phi : \pi_1(X, x_0) \to \pi(X)$ be the map given by "forgetting the basepoint".

Claim: Φ is onto. Proof: Let $f : S^1 \to X$ and let $f(1) = x_1$. Choose a path from x_0 to x_1 and let g be the loop at x_0 obtained by following this path from x_0 to x_1, then looping from x_1 to itself via f, and then following this path back from x_1 to x_0. A free homotopy between g and f is given by the following picture

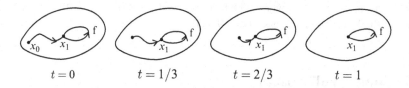

$$t = 0 \qquad\qquad t = 1/3 \qquad\qquad t = 2/3 \qquad\qquad t = 1$$

Claim: If g_1 and g_0 are conjugate elements of $\pi_1(X, x_0)$, then $f_1 = \Phi(g_1)$ and $f_0 = \Phi(g_0)$ are freely homotopic. *Proof:* Let $g_1 = h g_0 h^{-1}$ and let g_1 be represented by $\alpha : I \to X$, where I is parameterized by s, with α following h for $0 \le s \le 1/3$, α following g_0 for $1/3 \le s \le 2/3$, and α following h^{-1} for $2/3 \le s \le 1$. Let $A : I \times I \to X$ by $A(s, t) = \alpha(s + t/3)$ where $s + t/3$ is taken mod 1. Then A gives a free homotopy between g_1 and $\beta : I \to X$ with β following g_0 for $0 \le s \le 1/3$, β following h^{-1} for $1/3 \le s \le 2/3$, and β following h for $2/3 \le s \le 1$. But this map is obviously homotopic rel basepoint to g_0.

Claim: If $f_1 = \Phi(g_1)$ and $f_0 = \Phi(g_0)$ are freely homotopic, then g_1 and g_0 are conjugate elements of $\pi_1(X, x_0)$. *Proof:* Given a homotopy G between g_1 and g_0 we have a map of the square into X:

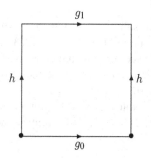

Then the left-hand side, the top, and the right-hand side give a loop that represents hg_1h^{-1}. But this loop is homotopic rel basepoint to g_0 by a homotopy that deforms those three sides of the square to the bottom as shown:

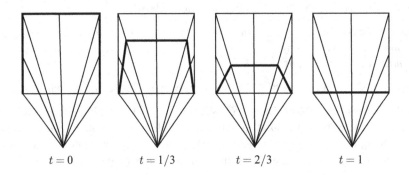

$$t = 0 \qquad\qquad t = 1/3 \qquad\qquad t = 2/3 \qquad\qquad t = 1$$

(That is, at time t one traverses the heavy curve from the lower left up, over, and down to the lower right as s goes from 0 to 1.)   □

2.6 Some "Bad" Spaces

Every mathematical theory is best equipped to handle some kinds of mathematical objects, which it considers to be "good", and less well equipped to handle others, which it considers to be "bad". Algebraic topology is no exception, and we will concentrate our attention on good spaces. But in this section we give a couple of examples of bad ones.

Example 2.6.1. Let X be the subspace of $\mathbb{R}^2$ consisting of the closed line segments joining the points $(1/n, 0)$ to $(0, 1)$ for each positive integer n, and also the closed line segment joining $(0, 0)$ to $(0, 1)$. Give X the topology it inherits as a subspace of $\mathbb{R}^2$. Then X is path connected but not locally path connected.

Let A be the subspace of X consisting of the closed line segment joining $(0, 0)$ to $(0, 1)$. Then X and A are both contractible to the point $(0, 1)$ by a homotopy leaving that point fixed. It then follows that A is a deformation retract of X. But A is not a strong deformation retract of X.   ◊

Example 2.6.2. Let H be the subset of $\mathbb{R}^2$ which consists of the union of circles of radius $1/n$ centered at the point $(1/n, 0)$, $n = 1, 2, 3, \ldots$. Give H the topology it inherits as a subset of $\mathbb{R}^2$. H is commonly known as the *Hawaiian earring*.

Let K_1 be the space which is the union of the closed line segments joining the points of H to the point $(0, 0, 1)$ in $\mathbb{R}^3$ (where we regard $\mathbb{R}^2 \subset \mathbb{R}^3$ as usual), topologized as a subset of $\mathbb{R}^3$. Note that K_1 is homotopic to the cone on H as defined in Definition 1.2.10. K_1 is path connected and simply connected (as it is contractible) but is neither locally simply connected nor semilocally simply connected. Let K_2 be the space obtained by reflecting K_1 through the origin, and let $K = K_1 \cup K_2$. Then $K_1 \cap K_2 = \{(0,0,0)\}$, so K_1 and K_2 are both simply connected and $K_1 \cap K_2$ is connected. If K_1 and K_2 were open subsets of K, we could apply van Kampen's theorem to conclude that K is simply connected. But it is not. For example, a loop starting at $(0,0,0)$ that alternately winds around infinitely many circles of ever-decreasing radii on the "right" and "left" sides of the origin, and winds up at the origin, is not null-homotopic. Indeed, this loop is not even freely homotopic to a constant loop. Also, although K is not simply connected, it has no covering space other than K itself, so in particular does not have a simply connected covering space. $\Diamond$

2.7 Exercises

Exercise 2.7.1. Prove Lemma 2.1.2.

Exercise 2.7.2. Prove Lemma 2.1.3.

Exercise 2.7.3. Prove Theorem 2.1.5.

Exercise 2.7.4. Give an example of a map $p : Y \to X$ satisfying the conditions in Lemma 2.2.2 which is *not* a covering projection.

Exercise 2.7.5. Verify that the maps in Example 2.2.3 are indeed covering projections.

Exercise 2.7.6. Let X be a simply connected space. Fix x_0, x_1, two arbitrary points at X. Let $f : I \to X$ and $g : I \to X$ with $f(0) = g(0) = x_0$ and $f(1) = g(1) = x_1$. Show that f and g are homotopic rel $\{0,1\}$.

Exercise 2.7.7. Prove directly the following special case of van Kampen's theorem: Let X be the union $X = X_1 \cup X_2$ of two open sets X_1 and X_2. Suppose that X_1, X_2, and $A = X_1 \cap X_2$ are all path-connected. Suppose that each of X_1 and X_2 is simply connected. Then X is simply connected.

Exercise 2.7.8. Prove parts (ii) and (iii) of Theorem 2.2.19.

Exercise 2.7.9. (a) Let C be a connected finite 1-complex. Show that the number of edges n of C not in a maximal tree T is well-defined, i.e., independent of choice of T.
(b) Show that C is homotopy equivalent to the n-leafed rose R_n for some n.

Exercise 2.7.10. Let p be a prime. Show that the free group $\langle a,b \rangle$ has exactly $p+1$ normal subgroups of index p.

Exercise 2.7.11. Find all 4-fold covers of the two-leafed rose, up to equivalence.

Exercise 2.7.12. Let X be a path-connected space and let $f : (X, x_0) \to (X, x_0)$ be a map. The mapping torus M_f of f is the identification space $X \times I / \sim$ where $(x, 0)$ is identified to $(f(x), 1)$. Let m_0 be the image of the point $(x_0, 0)$ in M_f. Show that $\pi_1(M_f, m_0) = \pi_1(X, x_0) * \pi_1(S^1)/tgt^{-1} = f_*(g)$, where t is the generator of $\pi_1(S^1)$.

Exercise 2.7.13. Let G be a topological group (*not* necessarily abelian) with identity element e. Let $\alpha \in \pi_1(G, e)$ be represented by $f : (I, \{0, 1\}) \to (G, e)$ and let $\beta \in \pi_1(G, e)$ be represented by $g : (I, \{0, 1\}) \to (G, e)$. Show that both $\alpha\beta$ and $\beta\alpha$ are represented by $h : (I, \{0, 1\}) \to (G, e)$ given by $h(t) = f(t)g(t)$. Note this implies that $\pi_1(G, e)$ is abelian, and also that α^{-1} is represented by $k : (I, \{0, 1\}) \to (G, e)$ given by $k(t) = (f(t))^{-1}$.

Exercise 2.7.14. Let $p(z) = a_n z^n + \cdots + a_0$ be a polynomial of degree $n > 0$ with complex coefficients. Let $q(z)$ be the polynomial $q(z) = z^n$. For a positive real number r, let $p_r(z) = p(rz)/\|p(rz)\|$ and similarly for $q_r(z)$. Show that, for r sufficiently large, $p_r : S^1 \to S^1$ and $q_r : S^1 \to S^1$ are homotopic.

Exercise 2.7.15. Prove the Fundamental Theorem of Algebra: Every nonconstant complex polynomial has a complex root.

Exercise 2.7.16. Prove the claims in Example 2.6.1.

Exercise 2.7.17. Prove the claims in Example 2.6.2.

Chapter 3
Generalized Homology Theory

We begin by presenting the famous Eilenberg-Steenrod axioms for homology. We then proceed by drawing (many) useful consequences from these axioms. At the end of this chapter we introduce cohomology.

We defer showing that there exist a nontrivial homology theory satisfying these axioms until later.

Throughout this chapter, unless otherwise stated, $H_i(X)$ and $H_i(X,A)$ denote generalized homology groups (as defined below), and all results are valid for generalized homology, not just ordinary homology.

3.1 The Eilenberg-Steenrod Axioms

A pair of spaces (X,A) is a topological space X and a subspace A. We identify the space X with the pair $(X,\emptyset)$.

Here are the Eilenberg-Steenrod axioms.

Definition 3.1.1. A *homology theory* associates to each pair (X,A) a sequence of abelian groups $\{H_i(X,A)\}_{i \in \mathbb{Z}}$ and a sequence of homomorphisms $\{\partial_i : H_i(X,A) \to H_{i-1}(A,\emptyset)\}_{i \in \mathbb{Z}}$ and to each map $f : (X,A) \to (Y,B)$ a sequence of homomorphisms $\{f_i : H_i(X,A) \to H_i(Y,B)\}_{i \in \mathbb{Z}}$ satisfying the following axioms, which hold for all values of i. (We abbreviate $H_i(X,\emptyset)$ to $H_i(X)$.)

Axiom 1. If $f : (X,A) \to (X,A)$ is the identity map, then $f_i : H_i(X,A) \to H_i(X,A)$ is the identity map.

Axiom 2. If $f : (X,A) \to (Y,B)$ and $g : (Y,B) \to (Z,C)$, and h is the composition $h = g \circ f$, $h : (X,A) \to (Y,C)$, then $h_i = g_i \circ f_i$.

Axiom 3. If $f : (X,A) \to (Y,B)$ then the following diagram commutes, where $f|A : A \to B$ is the restriction of f to A:

© Springer International Publishing Switzerland 2014
S.H. Weintraub, *Fundamentals of Algebraic Topology*, Graduate Texts in Mathematics 270, DOI 10.1007/978-1-4939-1844-7_3

$$H_i(X,A) \xrightarrow{\ f_i\ } H_i(Y,B)$$

$$\partial_i \downarrow \qquad\qquad \downarrow \partial_i$$

$$H_{i-1}(A) \xrightarrow{\ (f|A)_{i-1}\ } H_{i-1}(B)$$

Axiom 4. The homology sequence (where the first two maps are induced by the inclusions $A \hookrightarrow X$ and $(X,\emptyset) \hookrightarrow (X,A)$)

$$\cdots \longrightarrow H_i(A) \longrightarrow H_i(X) \longrightarrow H_i(X,A) \xrightarrow{\ \partial_i\ } H_{i-1}(A) \longrightarrow$$

is exact.

Axiom 5. If $f : (X,A) \to (Y,B)$ and $g : (X,A) \to (Y,B)$ are homotopic, then $f_i : H_i(X,A) \to H_i(Y,B)$ and $g_i : H_i(X,A) \to H_i(Y,B)$ are equal.

Axiom 6. If $U \subseteq A$ with the closure of U contained in the interior of A, and $f : (X-U, A-U) \to (X,A)$ is the inclusion, then

$$f_i : H_i(X-U, A-U) \longrightarrow H_i(X,A)$$

is an isomorphism.

Axiom 7. If X is a space consisting of a single point, then

$$H_i(X) = 0 \quad \text{for } i \neq 0. \qquad\qquad \diamond$$

Axiom 6 is known as excision, and it is convenient to reformulate it so as to obtain a definition.

A *couple* of subspaces (A,B) of a space X consists of two subspaces A and B of X, with no inclusion relations assumed.

Definition 3.1.2. Let (X,A) be a pair and let U be a subset of A. The inclusion $(X-U, A-U) \to (X,A)$ is an *excisive map* if it induces an isomorphism on homology.

If A and B are subspaces of X, then (A,B) is an *excisive couple* if the inclusion $(A, A\cap B) \to (A\cup B, B)$ is an excisive map. $\qquad \diamond$

With this definition, Axiom 6 states that if the closure of U is contained in the interior of A, then the inclusion $(X-U, A-U) \to (X,A)$ is an excisive map, or, equivalently, that in this situation $(X-U, A)$ is an excisive couple.

Axiom 7 is known as the dimension axiom, and the abelian group $G = H_0(*)$ is called the *coefficient group* of the homology theory.

Definition 3.1.3. A *generalized homology theory* is a theory satisfying Axioms 1 through 6. $\qquad \diamond$

When we wish to stress the difference between it and a generalized homology theory, we will refer to a homology theory as an *ordinary homology theory*.

Our results in this chapter will apply to any generalized homology theory, so in this chapter we let $H_i(X)$ or $H_i(X,A)$ denote a generalized homology group.

We are being careful by denoting the induced maps on homology by $f_i : H_i(X) \to H_i(Y)$, for example, but it is common practice to denote all these maps by f_*, so that $f_* : H_i(X) \to H_i(Y)$, and we shall sometimes follow this practice. (Sometimes the profusion of indices creates confusion rather than clarity.)

We now introduce, for later reference, a properly that a generalized homology theory may have.

Definition 3.1.4. A generalized homology theory is *compactly supported* if for any pair (X,A) and any element $\alpha \in H_i(X,A)$ there is a compact pair $(X_0,A_0) \subseteq (X,A)$ with $\alpha = j_*(\alpha_0)$ for some element $\alpha \in H_i(X_0,A_0)$, where $j : (X_0,A_0) \to (X,A)$ is the inclusion.   ◊

3.2 Consequences of the Axioms

We first list some consequences of the axioms. Again, i is allowed to be arbitrary.

Lemma 3.2.1.   (i) $H_i(\emptyset) = 0$.
(ii) *Let $f : (X,A) \to (Y,B)$ be a homotopy equivalence. Then $f_i : H_i(X,A) \to H_i(Y,B)$ is an isomorphism.*
(iii) *Suppose that A is a retract of X. Let $j : A \to X$ be the inclusion and $r : X \to A$ be a retraction. Then $j_i : H_i(A) \to H_i(X)$ is an injection and $r_i : H_i(X) \to H_i(A)$ is a surjection. Furthermore,*

$$H_i(X) \cong H_i(A) \oplus H_i(X,A)$$

and

$$H_i(X,A) \cong \operatorname{Ker}(r_i).$$

(iv) *Let X be a space and let X_1 and X_2 be unions of components of X. Let $j^1 : X_1 \to X$ and $j^2 : X_2 \to X$ be the inclusions. Then $j_i^1 + j_i^2 : H_i(X_1) \oplus H_i(X_2) \to H_i(X)$ is an isomorphism.*

Lemma 3.2.1(ii) has the following useful generalization

Theorem 3.2.2. *Let $f : (X,A) \to (Y,B)$ be a map of pairs and suppose that both $f : X \to Y$ and $f|A : A \to B$ are homotopy equivalences. Then $f_i : H_i(X,A) \to H_i(Y,B)$ is an isomorphism for all i.*

Proof. We have the commutative diagram of exact sequences:

$$\longrightarrow H_i(A) \longrightarrow H_i(X) \longrightarrow H_i(X,A) \longrightarrow H_{i-1}(A) \longrightarrow H_{i-1}(X) \longrightarrow$$
$$\downarrow \qquad\qquad \downarrow \qquad\qquad \downarrow \qquad\qquad \downarrow \qquad\qquad \downarrow$$
$$\longrightarrow H_i(B) \longrightarrow H_i(Y) \longrightarrow H_i(Y,B) \longrightarrow H_{i-1}(B) \longrightarrow H_{i-1}(Y) \longrightarrow$$

The first, second, fourth, and fifth vertical arrows are isomorphisms. Hence, by Lemma A.1.8, so is the third.   □

It is convenient to formulate the following:

Definition 3.2.3. Let X be a nonempty space. Let $f : X \to *$ (the space consisting of a single point). The *reduced* homology group .

$$\tilde{H}_i(X) = \mathrm{Ker}\,(f_i : H_i(X) \longrightarrow H_i(*)).$$
◇

Theorem 3.2.4. *Let X be a nonempty space and let x_0 be an arbitrary point of X. Then for each i,*

(i) $\tilde{H}_i(X) \cong H_i(X,x_0)$,
(ii) $H_i(X) \cong H_i(x_0) \oplus \tilde{H}_i(X)$.

Proof. As x_0 is a retract of X, this is a special case of Lemma 3.2.1.   □

Lemma 3.2.5. *Let $f : X \to Y$ be a map. Then f induces well-defined maps $\tilde{f}_i : \tilde{H}_i(X) \to \tilde{H}_i(Y)$ for every i, where $\tilde{f}_i = f_i | \tilde{H}_i(X)$.*

Proof. This follows immediately from the commutativity of the diagram

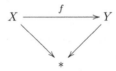

□

Remark 3.2.6. Note that $f : X \to Y$ does *not* induce $f_i : (X,x_0) \to (Y,y_0)$ unless $y_0 = f(x_0)$, so we must be careful with the isomorphisms in Theorem 3.2.4. If X and Y are both path connected the situation is not so bad. For if Z is a path-connected space and z_0 and z_1 are any two points in Z, the maps $j_0 : \{x\} \to \{z_0\} \subseteq Z$ and $j_1 : \{*\} \to \{z_1\} \subseteq Z$ are homotopic, so the maps of these on homology agree, i.e., $H_i(z_0)$ and $H_i(z_1)$ are the same subgroup of $H_i(Z)$. But if Z is not path connected, this need not be the case (see Remark 4.1.2).   ◇

We recall that if the closure of U is contained in the interior of A, then the inclusion $(X - U, A - U) \to (X,A)$ is excisive. It is easy to see that this cannot be weakened to the closure of U contained in A, and we give examples in

Example 4.1.11. It also cannot be weakened to U contained in the interior of A. Examples are harder to construct, but we give two in Example 5.2.8.

In practice, we often are in the situation where A is a closed set and we wish to excise the interior of A, i.e., we wish to have the inclusion $(X - \mathrm{int}(A), \partial A) \to (X, A)$ be excisive. This is true if ∂A sits "nicely" in X. To be precise, we have the following very useful result.

Theorem 3.2.7. (1) *Let A be a nonempty closed subset of X. Suppose that ∂A has an open neighborhood C in A such that the inclusions $(X - A) \cup C \to X - \mathrm{int}(A)$ and $\partial A \to C$ are both homotopy equivalences. Then $(X - \mathrm{int}(A), \partial A) \to (X, A)$ is excisive.*

(2) *Let A be a nonempty closed subset of X. Suppose that ∂A has an open neighborhood C in $X - \mathrm{int}(A)$ such that the inclusions $A \to C \cup A$ and $\partial A \to C$ are both homotopy equivalences. Then $(X - \mathrm{int}(A), \partial A) \to (X, A)$ is excisive.*

Proof. (1) Let $V = A - C$. Then V is a closed set in the interior of A, so $(X - V, A - V) \to (X, A)$ is excisive. But $X - V = (X - A) \cup C$ and $A - V = C$. By hypothesis the first of these is homotopy equivalent to $X - \mathrm{int}(A)$ and the second of these is homotopy equivalent to ∂A, so by Theorem 3.2.2 $(X - \mathrm{int}(A), \partial A) \to (X, A)$ is excisive.

(2) Note that $\mathrm{int}(A)$ is a set whose closure A is contained in the interior of $C \cup A$. Thus $(X - \mathrm{int}(A), (C \cup A) - \mathrm{int}(A)) \to (X, C \cup A)$. But $(C \cup A) - \mathrm{int}(A) = C$. By hypothesis C is homotopy equivalent to ∂A and $C \cup A$ to A, so again by Theorem 3.2.2 $(X - \mathrm{int}(A), \partial A) \to (X, A)$ is excisive. □

Remark 3.2.8. It is easiest to visualize this theorem by considering the following pictures, where in (1) C is a "collar" of ∂A inside A, and in (2) C is a "collar" of A outside A.

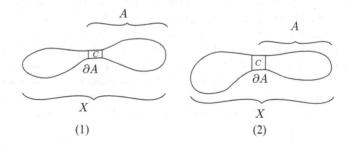

The way these situations most often arise is when when ∂A is a strong deformation retract of C. Note that in (2), when C is outside A, this is equivalent to $W = C \cup A$ being an open neighborhood of A having A as a strong deformation retract. ◊

This theorem allows us to give two interpretations of relative homology.

Recall we defined cA, the cone on A, in Definition 1.2.10. The identification space $X \cup_A cA$ is the quotient space of the disjoint union of X and cA under the identification of $a \in A \subseteq X$ with $(a,0) \in cA$. We let $*$ denote the "cone point", i.e., the point to which $A \times \{1\}$ is identified. Also, in the second half of the theorem, X/A is the quotient space of X obtained by identifying A to a point, and we let $*$ denote the point A/A.

Theorem 3.2.9. (1) *Let A be a nonempty subset of X. Then for each i, $H_i(X,A)$ is isomorphic to the reduced homology group $\tilde{H}_i(X \cup cA)$.*

(2) *Let A be a nonempty closed subset of X and suppose that A is a strong deformation retract of some neighborhood W of A. Thus for each i, $H_i(X,A)$ is isomorphic to the reduced homology group $\tilde{H}_i(X/A)$.*

Proof. (1) We follow the idea of the proof of Theorem 3.2.7. Let $V = \{(a,s) \mid s \geq \frac{1}{2}\}$ so that V is a closed subset of $X \cup_A cA$ which is contained in the interior of cA. Then the inclusion $((X \cup_A cA) - V, cA - V) \to (X \cup_A cA, cA)$ is excisive. But there is a strong deformation retraction of the pair $((X \cup_A cA) - V, cA - V)$ to (X,A), and a strong deformation retraction of cA to the point $*$, so we have an isomorphism on homology, for every i, from $H_i(X,A)$ to $H_i(X \cup_A cA, *)$. But the latter group is isomorphic to $\tilde{H}_i(X \cup_A cA)$.

(2) We have the following chain of isomorphisms, for each i:

$$H_i(X,A) \cong H_i(X,W)$$
$$\cong H_i(X - A, W - A)$$
$$\cong H_i(X/A - A/A, W/A - A/A)$$
$$\cong H_i(X/A, W/A)$$
$$\cong H_i(X/A, A/A) = H_i(X/A, *) \cong \tilde{H}_i(X/A).$$

The first isomorphism is because the inclusion $A \to W$ is a homotopy equivalence. The second isomorphism is ordinary excision (A is a closed subset of the open set W). The third isomorphism is because there is a homeomorphism of pairs, i.e., $X - A$ is homeomorphic to $X/A - A/A$, this homeomorphism restricting to a homeomorphism of $W - A$ to $W/A - A/A$. The fourth isomorphism is again ordinary excision, this time in the space X/A, as $* = A/A$ is closed and is in the open set W/A. By hypothesis, there is a strong deformation retraction from W into A in X, and this descends to give a strong deformation retraction from W/A to A/A in X/A, yielding the fifth isomorphism. $\square$

It is the excision property that enables us to actually compute homology groups. One of the most common ways it enters is the Mayer-Vietoris sequence.

Theorem 3.2.10 (Mayer-Vietoris). *Let $X = X_1 \cup X_2$, $A = X_1 \cap X_2$, and suppose that the inclusion $(X_1, A) \to (X, X_2)$ is excisive. Then there is a long exact sequence in homology*

$$\cdots \longrightarrow H_i(A) \xrightarrow{\alpha} H_i(X_1) \oplus H_i(X_2) \xrightarrow{\beta} H_i(X) \xrightarrow{\Delta} H_{i-1}(A) \longrightarrow \cdots .$$

Proof. We have the long exact homology sequences

$$
\begin{array}{ccccccccc}
\longrightarrow & H_i(A) & \xrightarrow{\alpha_1} & H_i(X_1) & \xrightarrow{\gamma_1} & H_i(X_1,A) & \xrightarrow{\partial_1} & H_{i-1}(A) & \longrightarrow \\
& \downarrow{\alpha_2} & & \downarrow{\beta_1} & & \downarrow{\varepsilon} & & \downarrow{\alpha_2} & \\
\longrightarrow & H_i(X_2) & \xrightarrow{\beta_2} & H_i(X) & \xrightarrow{\gamma_2} & H_i(X,X_2) & \xrightarrow{\partial_2} & H_{i-1}(X_2) & \longrightarrow
\end{array}
$$

where by assumption $\varepsilon : H_i(X_1,A) \to H_i(X,X_2)$ is an isomorphism. Then the theorem follows by applying Theorem A.2.11. □

A *triad* of spaces is a triple (X,A,B) with A and B each subspaces of X. In this situation we also have the following Mayer-Vietoris sequence.

Theorem 3.2.11. *Let (X,A,B) be a triad and suppose that the inclusion $(A,A \cap B) \to (A \cup B,B)$ is excisive. Then there is an exact homology sequence*

$$
\cdots \longrightarrow H_i(X,A \cap B) \longrightarrow H_i(X,A) \oplus H_i(X,B) \longrightarrow H_i(X,A \cup B)
$$
$$
\longrightarrow H_{i-1}(X,A \cap B) \longrightarrow \cdots .
$$

Proof. Exactly the same as the proof of Theorem 3.2.10. □

Here is a construction that we will be using later.

Definition 3.2.12. Let X be a space. The *suspension ΣX of X* is the quotient space of $X \times [-1,1]$ under an identification of $X \times \{1\}$ to a point c_+ and $X \times \{-1\}$ to a different point c_-. ◇

Note that the subspaces $c_+X = \{(x,s) \in \Sigma X \mid s \geqslant 0\}$ and $c_-X = \{(x,s) \in \Sigma X \mid s \leqslant 0\}$ are each homeomorphic to cX, the cone on X.

Also observe that any $f : X \to Y$ determines a map $\Sigma f : \Sigma X \to \Sigma Y$ defined by $\Sigma f(x,s) = (f(x),s)$, $x \in X$, $s \in [-1,1]$.

Theorem 3.2.13. (1) *For any space X there is an isomorphism, for any i,*

$$
\Sigma : \tilde{H}_{i+1}(\Sigma X) \longrightarrow \tilde{H}_i(X).
$$

(2) *For any $f : X \to Y$ the following diagram commutes:*

$$
\begin{array}{ccc}
\tilde{H}_{i+1}(\Sigma X) & \xrightarrow{\Sigma} & \tilde{H}_i(X) \\
\downarrow{(\Sigma f)_{i+1}} & & \downarrow{f_i} \\
\tilde{H}_{i+1}(\Sigma Y) & \xrightarrow{\Sigma} & \tilde{H}_i(Y).
\end{array}
$$

Proof. We prove (1). We have the exact homology sequence of the pair (c_+X, X):

$$\cdots \longrightarrow H_{i+1}(c_+X, X) \longrightarrow H_i(X) \longrightarrow H_i(c_+X) \longrightarrow \cdots$$

Now c_+X is contractible to the cone point c_+, and furthermore there is a path in c_+X from c_+ to any point x_0 of X. This implies that the inclusion of x_0 into X gives a splitting of the map $H_i(X) \to H_i(c_+X)$, so this map is a surjection and the exact homology sequence breaks up into a sequence of (split) short exact sequences

$$0 \longrightarrow H_{i+1}(c_+X, X) \longrightarrow H_i(X) \longrightarrow H_i(*) \longrightarrow 0$$

and so $H_{i+1}(c_+X, X)$ is the kernel of the map $H_i(X) \to H_i(*)$, which by definition is $\tilde{H}_i(X)$. But then by Theorem 3.2.9(2) $H_{i+1}(c_+X, X) = H_{i+1}(c_+X/X, X/X)$. But this pair is homeomorphic to $(\Sigma X, c_-)$ (under the homeomorphism $(x, s) \to (x, 2s - 1)$). Thus $H_{i+1}(c_+X, X) \cong H_{i+1}(\Sigma X, c_-) \cong \tilde{H}_{i+1}(\Sigma X)$ by Theorem 3.2.4. $\square$

The exact homology sequence of a pair has a generalization to a triple.

Theorem 3.2.14. *Let A and B be subspaces of X with $B \subseteq A$. Then there is an exact homology sequence*

$$\cdots \longrightarrow H_i(A, B) \longrightarrow H_i(X, B) \longrightarrow H_i(X, A) \xrightarrow{\partial} H_{i-1}(A, B) \longrightarrow \cdots.$$

Proof. We merely remark here that the boundary map in the sequence is the composition

$$H_i(X, A) \longrightarrow H_{i-1}(A) \longrightarrow H_{i-1}(A, B).$$

Otherwise, the result follows directly from Theorem A.2.12. $\square$

Once we have the exact sequence of a triple, the proof of the Mayer-Vietoris sequence goes through unchanged to give the following result.

Theorem 3.2.15. *Let $X = X_1 \cup X_2$, $A = X_1 \cap X_2$, and suppose that the inclusion $(X_1, A) \to (X, X_2)$ is excisive. Let B be an arbitrary subspace of A. Then there is a long exact sequence in homology*

$$\cdots \to H_i(A, B) \to H_i(X_1, B) \oplus H_i(X_2, B) \to H_i(X, B) \to H_{i-1}(A, B) \to \cdots.$$

3.3   Axioms for Cohomology and Their Consequences

We now present the Eilenberg-Steenrod axioms for cohomology theory, and develop the theory axiomatically. The development closely parallels that of homology theory "with all the arrows reversed," so we shall be brief.

Definition 3.3.1. A *cohomology theory* associates to each pair (X,A) a sequence of abelian groups $\{H^i(X,A)\}_{i\in\mathbb{Z}}$ and a sequence of group homomorphisms $\{\delta^i : H^i(A,\emptyset) \to H^{i+1}(X,A)\}_{i\in\mathbb{Z}}$ and to each map $f : (X,A) \to (Y,B)$ a sequence of homomorphisms $\{f^i : H^i(Y,B) \to H^i(X,A)\}_{i\in\mathbb{Z}}$ satisfying the following axioms, which hold for all values of i. (We abbreviate $H^i(X,\emptyset)$ to $H^i(X)$.)

Axiom 1. If $f : (X,A) \to (X,A)$ is the identity map, then $f^i : H^i(X,A) \to H^i(X,A)$ is the identity map.

Axiom 2. If $f : (X,A) \to (X,B)$ and $g : (Y,B) \to (Z,C)$ and h is the composition $h = g \circ f, h : (X,A) \to (Y,C)$, then $h^i = f^i \circ g^i$.

Axiom 3. If $f : (X,A) \to (Y,B)$ then the following diagram commutes, where $f|A : A \to B$ is the restriction of f to A:

$$
\begin{array}{ccc}
H^i(X,A) & \xleftarrow{\;f^i\;} & H^i(X,B) \\
{\scriptstyle \delta^{i-1}}\big\uparrow & & \big\uparrow{\scriptstyle \delta^{i-1}} \\
H^{i-1}(A) & \xleftarrow{(f|A)^{i-1}} & H^{i-1}(B).
\end{array}
$$

Axiom 4. The cohomology sequence

$$\cdots \longleftarrow H^i(A) \longleftarrow H^i(X) \longleftarrow H^i(X,A) \longleftarrow H^{i-1}(A) \longleftarrow \cdots$$

is exact.

Axiom 5. If $f : (X,A) \to (Y,B)$ and $g : (X,A) \to (Y,B)$ are homotopic, then $f^i : H^i(Y,B) \to H^i(X,A)$ and $g^i : H^i(Y,B) \to H^i(X,A)$ are equal.

Axiom 6. If $U \subseteq A$ with the closure of U contained in the interior of A, and $f : (X-U,A-U) \to (X,A)$ is the inclusion, then

$$f^i : H^i(X,A) \longrightarrow H^i(X-U,A-U)$$

is an isomorphism.

Axiom 7. If X is a space consisting of a single point, then

$$H^i(X) = 0 \quad \text{for } i \neq 0. \qquad \qquad \diamond$$

We invite the reader to revisit the previous section and derive the analogs in cohomology for the results stated there for homology. We shall merely state here a few salient points.

Definition 3.3.2. Let X be a nonempty space. Let $f : X \to *$ (the space consisting of a single point). The *reduced* cohomology group

$$\tilde{H}^i(X) = \operatorname{Coker}\left(f^i : H^i(*) \longrightarrow H^i(X)\right).$$
                                                                                                          $\Diamond$

Remark 3.3.3. Note that $\tilde{H}_i(X)$ is a subgroup of $H_i(X)$ while $\tilde{H}^i(X)$ is a quotient of $H^i(X)$.
                                                                                                          $\Diamond$

We have the Mayer-Vietoris sequence in cohomology.

Theorem 3.3.4. *Let* $X = X_1 \cup X_2$, $A = X_1 \cap X_2$, *and suppose that the inclusion* $(X_1, A) \to (X, X_2)$ *is excisive. Then there is a long exact sequence in cohomology*

$$\cdots \longleftarrow H^i(A) \longleftarrow H^i(X_1) \oplus H^i(X_2) \longleftarrow H^i(X) \longleftarrow H^{i-1}(A) \longleftarrow \cdots.$$

3.4 Exercises

Exercise 3.4.1. Prove Lemma 3.2.1.

Exercise 3.4.2. Show that Axiom 6 (excision) is equivalent to: If A and B are subspaces of X with $X = \operatorname{int}(A) \cup \operatorname{int}(B)$, and $g : (A, A \cap B) \to (X, B)$ is the inclusion, then

$$g_i : H_i(A, A \cap B) \longrightarrow H_i(X, B)$$

is an isomorphism for each i.
    In other words, the following two conditions are equivalent:

(a) $(X - U, A)$ is an excisive couple whenever closure $(U) \subseteq$ interior (A).
(b) (A, B) is an excisive couple whenever $X = \operatorname{int}(A) \cup \operatorname{int}(B)$.

Exercise 3.4.3. Prove the observations in Remark 3.2.8:

(1) If ∂A has an open neighborhood C in A with ∂A a strong deformation retract of C, then the hypotheses of Theorem 3.2.7(1) are satisfied.
(2a) If ∂A has an open neighborhood C in $X - \operatorname{int}(A)$ with ∂A a strong deformation retract of C, then the hypotheses of Theorem 3.2.7(2) are satisfied.
(2b) The hypothesis of (2a) is satisfied if and only if $W = C \cup A$ is an open neighborhood of A having A as a strong deformation retract.

Exercise 3.4.4. Prove Theorem 3.2.13(2).

Exercise 3.4.5. Prove that for any i and n, $\tilde{H}_i(S^n)$ is isomorphic to $H_{i-n}(*)$.

Exercise 3.4.6. Let A be a path-connected subspace of the path connected space X. Show there is a long exact homology sequence

$$\cdots \longrightarrow \tilde{H}_i(A) \longrightarrow \tilde{H}_i(X) \longrightarrow H_i(X, A) \longrightarrow \tilde{H}_{i-1}(A) \longrightarrow \tilde{H}_{i-1}(X) \longrightarrow \cdots.$$

Exercise 3.4.7. Let X be a nonempty space and let $f : X \to Y$ be a map. The *mapping cone* C_f of f is the quotient space $X \times I \cup Y / \sim$ where $X \times \{0\}$ is identified to a single point and $(x, 1)$ is identified to $f(x)$, for each $x \in X$. Show there is a long exact homology sequence

$$\cdots H_i(X) \longrightarrow H_i(Y) \longrightarrow H_i(C_f) \longrightarrow H_{i-1}(X) \longrightarrow H_{i-1}(Y) \longrightarrow \cdots .$$

Exercise 3.4.8. Let X be a space and x_0 a point of X. The reduced suspension $\Sigma_0 X$ of X (at x_0) is the quotient space of ΣX under the further relation that (x_0, t) is identified with $(x_0, 0)$ for every $t \in [-1, 1]$. Show that the quotient map $q : \Sigma X \to \Sigma_0 X$ induces isomorphisms $q_i : H_i(\Sigma X) \to H_i(\Sigma_0 X)$ for every i.

Exercise 3.4.9. Formulate and prove the analog of Lemma 3.2.1 for cohomology.

Exercise 3.4.10. Show that the proofs of Theorems 3.2.7 and 3.2.9 go through unchanged to yield analogous results for cohomology.

Exercise 3.4.11. Formulate and prove the analog of Theorem 3.2.13 for cohomology.

Chapter 4
Ordinary Homology Theory

In this chapter we continue to proceed axiomatically. We assume now that we have an ordinary homology theory, i.e., one that satisfies the dimension axiom, and we assume in addition that the coefficient group is the integers $\mathbb{Z}$. Throughout this chapter $H_n(X)$, or $H_n(X,A)$, will denote such a homology group.

There is one thing we need to be precise about. We have that $H_0(*) \cong \mathbb{Z}$, where $*$ is the space consisting of a single point. We now choose, once and for all, a space $*$ consisting of a single point and an isomorphism of $H_0(*)$ with $\mathbb{Z}$. Given this isomorphism we identify $H_0(*)$ with $\mathbb{Z}$. We use the notation 1_* to denote the generator of $H_0(*)$ that we identify with the generator $1 \in \mathbb{Z}$. In addition, if p is any other space consisting of a single point, we have a unique map $f : * \to p$ including an isomorphism $f_0 : H_0(*) \to H_0(p)$, and we let $1_p = f_0(1_*)$.

4.1 Homology Groups of Spheres, and Some Classical Applications

In this section we compute homology groups of spheres, and related matters. Given our previous work, the computation is easy, but we will have to be careful. We then use our results to easily derive some classical, and important, applications.

For ease of notation we set $g = 1_* \in H_0(*)$.

Let $X = S^0 = \{-1, 1\}$. Then the map $j : * \to -1$ induces a map $j_0 : H_0(*) \to H_0(S^0)$ and we set $q = j_0(g)$. Also, the map $k : * \to 1$ induces a map $k_0 : H_0(*) \to H_0(S^0)$ and we set $p = k_0(g)$. We let $r : S^0 \to *$ be the unique map.

Lemma 4.1.1. *1. $H_0(S^0) \cong \mathbb{Z} \oplus \mathbb{Z}$. More precisely, $H_0(S^0) = \{mp + nq \mid m, n \in \mathbb{Z}\}$.*
2. $\tilde{H}_0(S^0) \cong \mathbb{Z}$. More precisely, $\tilde{H}_0(S^0) = \{n(q - p) \mid n \in \mathbb{Z}\}$.
3. $\tilde{H}_i(S^0) = H_i(S^0) = 0$ for $i \neq 0$.

© Springer International Publishing Switzerland 2014
S.H. Weintraub, *Fundamentals of Algebraic Topology*, Graduate Texts in Mathematics 270, DOI 10.1007/978-1-4939-1844-7_4

Proof. (1) Since $\{-1\}$ and $\{1\}$ are distinct components of S^0, we have by Lemma 3.2.1 that $H_i(S^0) \cong H_i(\{-1\}) \oplus H_i(\{1\})$. Since $\{-1\}$ and $\{1\}$ are both spaces consisting of a single point, the maps j and k are isomorphisms on homology. Now $H_i(*) = 0$ for $i \neq 0$, so $H_i(S^0) = 0$ for $i \neq 0$, and as $\tilde{H}_i(S^0)$ is subgroup of $H_i(S^0)$, $\tilde{H}_i(S^0) = 0$ for $i \neq 0$ as well.

More interestingly, $j_0 : H_0(\{*\}) \to H_0(\{-1\})$ and $k_0 : H_0(\{*\}) \to H_0(\{1\})$ are isomorphisms, so $H_0(\{-1\}) \cong \mathbb{Z}$ is generated by q and $H_0(\{1\}) \cong \mathbb{Z}$ is generated by p.

Also, $\tilde{H}_0(S^0) = \text{Ker}(r_0 : H_0(S^0) \to H_0(\{*\}))$. Now the composition $\{*\} \to \{-1\} \to \{*\}$ is the identity map so $r_0(q) = g$ and the composition $\{*\} \to \{1\} \to \{*\}$ is also the identity map so $r_0(p) = g$ as well.

Thus $\text{Ker}(r_0)$ is generated by $q - p$.   □

Remark 4.1.2. Note that we have the excision isomorphism induced by the inclusion $(\{1\}, \emptyset) \to (S^0, \{-1\})$ and via this isomorphism we obtain $H_0(S^0, \{-1\}) \cong \mathbb{Z}$ is the subgroup generated by p. Similarly, $(\{-1\}, \emptyset) \to (S^0, \{1\})$ is excisive, giving the subgroup $H_0(S^0, \{1\}) \cong \mathbb{Z}$ generated by q. Also, we have seen that $\tilde{H}_0(S^0) \cong \mathbb{Z}$, generated by $q - p$. Thus these three subgroups, while isomorphic, are not identical.   ◇

Lemma 4.1.3. *Fix a positive integer n.*

1. $H_n(S^n) \cong \mathbb{Z}$ and $H_0(S^n) \cong \mathbb{Z}$.
2. $\tilde{H}_n(S^n) \cong \mathbb{Z}$.
3. $H_i(S^n) = 0$ for $i \neq 0, n$ and $\tilde{H}_i(S^n) = 0$ for $i \neq n$.

Proof. Observe that for any k, ΣS^k is homeomorphic to S^{k+1}. Then, if Σ^i denotes Σ applied i times, $\Sigma^i S^k$ is homeomorphic to S^{k+i}. In particular $\Sigma^n S^0$ is homeomorphic to S^n. Then, by repeated applications of Theorem 3.2.13 (or, more properly, by induction) and by Lemma 4.1.1, $\tilde{H}_i(S^n) \cong \mathbb{Z}$ for $i = n$ and 0 otherwise. The rest of the lemma follows easily.   □

Corollary 4.1.4. *Fix a positive integer n. Then $H_i(D^n, S^{n-1}) \cong \mathbb{Z}$ for $i = n$ and 0 for $i \neq n$.*

Lemma 4.1.5. *For any $n \geqslant 1$, there does not exist a retraction from D^n onto S^{n-1}.*

Proof. If there were such a retraction $r : D^n \to S^{n-1}$, then r would induce a surjection $r_i : H_i(D^n) \to H_i(S^{n-1})$ for each i, by Lemma 3.2.1(iii).

But for $i = n - 1$, $H_{n-1}(D^n) = 0$ and $H_{n-1}(S^{n-1}) = \mathbb{Z}$, so this is impossible.   □

Theorem 4.1.6 (Brouwer fixed-point theorem). *Let $f : D^n \to D^n$ be an arbitrary map. Then f has a fixed point, i.e. there is an $x_0 \in D$ with $f(x_0) = x_0$.*

Proof. Suppose that f does not have a fixed point. Let $r : D^n \to S^{n-1}$ be the map defined as follows:

For $x \in D^n$, take the line segment from $f(x)$ to x and prolong it until it intersects S^{n-1} at some point x'. Then set $r(x) = x'$. The map r is a retraction from D^n onto S^{n-1}. But no such map can exist, by Lemma 4.1.5.   □

Theorem 4.1.7 (Invariance of domain). *Let U be a nonempty open set in $\mathbb{R}^n$ and V be a nonempty open set in $\mathbb{R}^m$ and suppose there is a homeomorphism $f : U \to V$. Then $m = n$.*

Proof. This is trivially true if $m = 0$ or $n = 0$, so we assume $m \geqslant 1$ and $n \geqslant 1$.

Although from a logical standpoint it is not necessary to begin with this special case, the basic idea of the proof comes through most clearly if we first consider the case $U = \mathbb{R}^m$, $V = \mathbb{R}^n$. Thus suppose we have a homeomorphism $f : \mathbb{R}^m \to \mathbb{R}^n$. Let $y \in \mathbb{R}^n$ be arbitrary and let $x = f^{-1}(y)$. Then we have a homeomorphism of pairs $f : (\mathbb{R}^m, \mathbb{R}^m - \{x\}) \to (\mathbb{R}^n, \mathbb{R}^n - \{y\})$ which then induces an isomorphism on homology $f_i : H_i(\mathbb{R}^m, \mathbb{R}^m - \{x\}) \to H_i(\mathbb{R}^n, \mathbb{R}^n - \{y\})$ for each i. But, for $k \geqslant 1$, $H_i(\mathbb{R}^k, \mathbb{R}^k - \{x\}) = 0$ for $i \neq k - 1$ and $= \mathbb{Z}$ for $i = k - 1$, so we must have $m - 1 = n - 1$ and hence $m = n$.

Now we prove the general case.

Choose a point $y \in V$ and let $x = f^{-1}(y)$. For some $\varepsilon > 0$, V contains a ball N of radius ε around y. Let $M = f^{-1}(N)$, so $F : M \to N$ is a homeomorphism. Now for some $\delta > 0$, M contains a ball B of radius δ around x. Let $C = f(B)$, so that $f : B \to C$ is a homeomorphism. Then $f : (B, B - x) \to (C, C - y)$ is a homeomorphism of pairs. On the one hand, $H_i(B, B - x) \cong H_i(\mathbb{R}^m, \mathbb{R}^m - x)$ for every i. On the other hand, the closure of $N - C$ is contained in the interior of $C - y$, so by excision $H_i(C, C - y)$ is isomorphic to $H_i(N, N - y) \cong H_i(\mathbb{R}^n, \mathbb{R}^n - y)$ for every i, so as in the special case, $m = n$ and we are done. $\qquad\square$

Definition 4.1.8. Let $f : S^n \to S^n$ be a map. Then f induces $f_n : H_n(S^n) \to H_n(S^n)$. Since $H_n(S^n) \cong \mathbb{Z}$, f_n is multiplication by some integer d. This integer d is the *degree* of the map. $\qquad\diamond$

We shall need the following result when we investigate real projective spaces.

Lemma 4.1.9. *Let $a : S^n \to S^n$ be the antipodal map, i.e., $a(x_1, \ldots, x_{n+1}) = (-x_1, \ldots, -x_{n+1})$. Then the degree of a is $(-1)^{n+1}$.*

Proof. We divide the proof into two cases.

Case 1 (n is odd, $n = 2m - 1$). Then we may regard S^n as the unit sphere in $\mathbb{C}^m$, and $a : S^n \to S^n$ is $a(z_1, \ldots, z_m) = (-z_1, \ldots, -z_m)$. But there is a homotopy between the identity map on S^n and a given by $f_t(z_1, \ldots, z_m) = e^{\pi i t}(z_1, \ldots, z_m)$ so a induces the identity map on (reduced) homology.

Case 2 (n is even). First consider the case $n = 0$. Then a is the map which interchanges the points -1 and 1 of S^0. Hence the induced map on $H_0(S^0)$ takes q to p and p to q. In particular it takes the class $q - p$ to the class $p - q$, so by Lemma 4.1.1 it is multiplication by -1 on $\tilde{H}_0(S^0)$.

Now let $n = 2m > 0$. Then we may regard S^n as the unit sphere in $\mathbb{R} \times \mathbb{C}^m$, and $a : S^n \to S^n$ by $a(x, z_1, \ldots, z_m) = (-x, -z_1, \ldots, -z_m)$. Let $a' : S^n \to S^n$ by $a'(x, z_1, \ldots, z_m) = (-x, z_1, \ldots, z_m)$. Then there is a homotopy between a' and a given by $f_t(x, z_1, \ldots, z_m) = \left(-x, e^{\pi i t}(z_1, \ldots, z_m)\right)$. But a' is just $\Sigma^n \alpha$, where α is the antipodal map on S^0, so $\deg(a) = \deg(a') = \deg(\alpha) = -1$ by the $n = 0$ case. $\qquad\square$

Remark 4.1.10. We have many homology groups isomorphic to $\mathbb{Z}$. Since $\mathbb{Z}$ has two generators, there are choices of isomorphisms to be made, i.e, choices of generators. We now observe that we can make these choices consistently. We shall call the generators we choose the *standard generators*. We proceed inductively.

We choose the generator $\tilde{\sigma}_0 = q - p$ of $\tilde{H}_0(S^0)$, in the notation of Lemma 4.1.1.

Now suppose that, for $n \geqslant 1$, we have chosen $\tilde{\sigma}_{n-1}$, a generator of $\tilde{H}_{n-1}(S^{n-1})$. We have $\partial : H_n(D^n, S^{n-1}) \to \tilde{H}_{n-1}(S^{n-1})$ an isomorphism, and we let $\delta_n = \partial^{-1}(\tilde{\sigma}_{n-1})$. We have a homeomorphism of pairs $f : (D^n_*, S^{n-1}_0) \to (D^n, S^{n-1})$ by $f\left(t_1, \ldots, t_n, \sqrt{1 - (t_1^2 + \cdots + t_n^2)}\right) = (t_1, \ldots, t_n)$ and we let $\delta_n^+ = (f_*)^{-1}(\delta_n)$. We have maps

$$S^n \longrightarrow (S^n, D^n_-) \longleftarrow (D^n_+, S^{n-1}_0)$$

inducing isomorphisms on homology

$$H_n(S^n) \longrightarrow H_n(S^n, D^n_-) \longleftarrow H_n(D^n_+, S^{n-1}_0),$$

the right-hand map being an isomorphism by excision, and we let σ_n be the class in $H_n(S^n)$ whose image is δ_n^+. Finally, for $n \geqslant 1$, $\tilde{H}_n(S^n) \to H_n(S^n)$ is an isomorphism on homology and we let $\tilde{\sigma}_n$ be the inverse image of σ_n. ◇

We now give a pair of examples to show that the condition $\text{closure}(U) \subseteq \text{interior}(A)$ for $(X - U, A - U) \to (X, A)$ to be excisive cannot in general be weakened to $\text{closure}(u) \subseteq A$.

Example 4.1.11. (a) Let $X = [0, 1]$ and $A = \{1\}$. Let $U = \{1\}$, a closed set.

Then $U \subseteq A$ but the closure of U (i.e., U itself) is not contained in the interior of A. Now $(X - U, A - U) = ([0, 1), \emptyset)$ and $[0, 1)$ is homotopy equivalent to a point, so in particular $H_0(X - U, A - U) = \mathbb{Z}$. On the other hand, the inclusion $A \to X$ is a homotopy equivalence so it induces an isomorphism on homology groups, and so in particular $H_0(X, A) = 0$. Thus $(X - U, A - U) \to (X, A)$ is not excisive.

(b) Here is a more interesting example along the same lines.

Fix $n \geqslant 1$ and let $X = S^n$. Let $*$ denote an arbitrary point of S^n and let $U = A = \{*\}$.

As we have seen, $H_n(S^n) = \mathbb{Z}$, which readily implies $H^n(S^n, *) = \mathbb{Z}$. But $H^n(S^n - *, * - *) = H^n(S^n - *) = 0$ as $S^n - *$ is contractible.

◇

4.2 CW-Complexes and Cellular Homology

In this section we introduce a class of spaces, CW-complexes, that are particularly amenable to the methods of algebraic topology, as well as a kind of homology theory, cellular homology, that is particularly useful in studying them.

Definition 4.2.1. A space X is obtained from a subspace A by *adjoining an n-cell* if there is a map $f : S^{n-1} \to A$, called the *characteristic map* or *attaching map* of the n-cell, with X the identification space $X = A \cup D^n / \sim$ where $\sim$ is the identification $p \in S^{n-1} \sim f(p) \in A$, with the quotient topology.   ◊

Remark 4.2.2. Observe that the obvious map $D^n \to X$ restricts to a homeomorphism from $\mathrm{int}\,(D^n)$ onto its image in X.   ◊

Remark 4.2.3. Observe that if $n = 0$, then X is just the disjoint union of A and an isolated point.   ◊

Definition 4.2.4. A CW-*structure* on a space X is a union of subspaces

$$\emptyset = X^{-1} \subseteq X^0 \subseteq X^1 \subseteq X^2 \subseteq \cdots \subseteq X$$

such that, for each n, X^n is obtained from X^{n-1} by adjoining n-cells, i.e., for some indexing set Λ_n, and for each $\lambda \in \Lambda_n$, there are maps $f_\lambda : S_\lambda^{n-1} \to X^{n-1}$ with

$$X^n = X^{n-1} \cup \bigcup_{\lambda \in \Lambda_n} D_\lambda^n / \sim$$

where $\sim$ is the identification $p \in S_\lambda^{n-1} \sim f(p) \in X^{n-1}$, with X^n having the quotient topology, and furthermore

1. $X = \bigcup_{n=0}^\infty X^n$.
2. X has the weak topology with respect to $\{X^n\}$, i.e., $A \subseteq X$ is closed if $A \cap X^n$ is closed in X^n for every n.

A space X with a CW-structure is called a CW-*complex*. If X is a CW-complex and A is a subcomplex, the (X,A) is a CW-*pair*.   ◊

Definition 4.2.5. The image of $\mathrm{int}\,(D_\lambda^n)$ in X is a *cell*, or, more precisely, an n-*cell*.   ◊

Remark 4.2.6. Observe that for $\lambda \in \Lambda_n$ the inclusion of $\mathrm{int}\,(D_\lambda^n)$ into X is a homeomorphism onto its image. Also observe that if $\lambda, \mu \in \Lambda_n$, $\lambda \neq \mu$, then the image of $\mathrm{int}\,(D^n)$ and $\mathrm{int}\,(D_\mu^n)$ under these inclusions are disjoint.
Also, as a set, X is the disjoint union of its cells.   ◊

Lemma 4.2.7. *A CW-complex X has the following properties:*

1. *(Closure-finiteness) The closure of each cell in X intersects only finitely many other cells in X.*
2. *(Weak topology) A subset A of X is closed if and only if the intersection of A with the closure of every cell in X is closed in X.*

Proof. (1) The closure of each cell is $f(D_\lambda^n)$, the image of a compact set, and hence compact, and if (1) were false this set would have an infinite subset (one point from each other cell) without an accumulation point, which is impossible.   □

Definition 4.2.8. If $X = X^n$ but $X \neq X^{n-1}$ then X is *n-dimensional*. If X has only finitely many cells, then X is a *finite complex*.   ◊

Example 4.2.9. 1. A discrete set of points is a 0-dimensional CW-complex.
2. A 1-complex as defined in Definition 2.4.1 is a 1-dimensional CW-complex.
3. For $n \geqslant 1$, S^n has the structure of a CW-complex with one cell in dimension 0 and one cell in dimension n.
4. Consider the inclusions $S^0 \subset S^1 \subset S^2 \subset \cdots$ where S^{n-1} is the "equator" in S^n, separating S^n into two "hemispheres". This gives S^n the structure of a CW-complex with two cells in each dimension from 0 to n.
5. For $n \geqslant 2$, begin with the CW-structure on S^{n-1}, the equator, with one cell in dimension 0 and one cell in dimension $n-1$, and obtain S^n by attaching two n-cells, the northern and southern hemispheres.   ◊

Lemma 4.2.10. *Let X be obtained from A by adjoining an n-cell. Then*

$$H_i(X,A) = \begin{cases} \mathbb{Z} & i = n \\ 0 & i \neq n. \end{cases}$$

Proof. Let $C = \{x \in D^n \mid |x| \geqslant 1/2\}$. Then C is a "collar" of A as in Remark 3.2.8, i.e., A is a strong deformation retract of $A \cup C$. Thus $H_i(X,A) \cong H_i(D^n,C) \cong H_i(D^n,S^{n-1})$.   □

Lemma 4.2.11. *1. $H_i(X^n,X^{n-1}) = 0$ for $i \neq n$.*
2. For each $\lambda \in \Lambda_n$, $f_\lambda : (D^n_\lambda, S^{n-1}_\lambda) \to (X^n, X^{n-1})$ induces monomorphisms on homology, and furthermore

$$H_n(X^n,X^{n-1}) = \bigoplus_{\lambda \in \Lambda_n} (f_\lambda)_* \left(H_n(D^n_\lambda, S^{n-1}_\lambda) \right).$$

3. $H_i(X^n) = 0$ for $i > n$.
4. The inclusion $X^{n-1} \to X^n$ induces maps $H_i(X^{n-1}) \to H_i(X^n)$ that are isomorphisms except possibly for $i = n-1, n$.
5. There is an exact sequence

$$0 \longrightarrow H_n(X^n) \longrightarrow H_n(X^n,X^{n-1}) \longrightarrow H_{n-1}(X^{n-1}) \longrightarrow H_{n-1}(X^n) \longrightarrow 0.$$

Proof. This is just an elaboration of Lemma 4.2.10.

Let $(D^n(\frac{1}{2}), S^{n-1}(\frac{1}{2}))$ be the pair consisting of the disk of radius $\frac{1}{2}$ and its boundary. Then the inclusions induce isomorphisms on homology

$$H_* \left(D^n\left(\frac{1}{2}\right), S^{n-1}\left(\frac{1}{2}\right) \right) \longrightarrow H_* \left(D^n, \overline{D^n - D^n\left(\frac{1}{2}\right)} \right) \longrightarrow H_*(D^n, S^{n-1}).$$

Let $X' = X^{n-1} \cup \bigcup_\lambda f_\lambda (\overline{D^n - D^n(\frac{1}{2})}) \subseteq X^n$. Note that X^{n-1} is closed and contained in the interior of X', and that X^{n-1} is a strong deformation retract of X'.

Then we have a string of isomorphisms

$$\bigoplus_\lambda H_*(D_\lambda^n, S_\lambda^{n-1}) \cong \bigoplus_\lambda H_*\left(D_\lambda^n\left(\frac{1}{2}\right), S_\lambda^{n-1}\left(\frac{1}{2}\right)\right)$$

$$\cong \bigoplus_\lambda H_*\left(\text{int}(D_\lambda^n), \overline{\text{int}(D_\lambda^n) - D_\lambda^n\left(\frac{1}{2}\right)}\right)$$

$$\cong H_*(X^n - X^{n-1}, X' - X^{n-1})$$

$$\cong H_*(X^n, X') \cong H_*(X^n, X^{n-1})$$

where the fourth isomorphism is excision.

This immediately gives (1) and (2). Then (3) follows by induction, and (4) and (5) follow from the exact sequence of the pair (X^n, X^{n-1}). □

Corollary 4.2.12. *1. Let $m > n$. Then the inclusion $X^n \hookrightarrow X^m$ induces isomorphisms $H_i(X^n) \to H_i(X^m)$ for $i < n$ and an epimorphism $H_n(X^n) \to H_n(X^m)$.*

2. If X is finite dimensional or H_* is compactly supported, the inclusion $X^n \hookrightarrow X$ induces isomorphisms $H_i(X^n) \to H_i(X)$ for $i < n$ and an epimorphism $H_n(X^n) \to H_n(X)$.

Proof. (2) Any compact subset of X is contained in X^m for some m (compare the proof of Lemma 4.2.7). □

With this lemma in hand, we now define cellular homology.

Definition 4.2.13. Let X be a CW-complex. The *cellular chain complex* of X is defined by

$$C_n^{\text{cell}}(X) = H_n(X^n, X^{n-1})$$

with $\partial_n : C_n^{\text{cell}}(X) \to C_{n-1}^{\text{cell}}(X)$ the composition

$$H_n(X^n, X^{n-1}) \xrightarrow{\partial} H_{n-1}(X^{n-1}) \longrightarrow H_{n-1}(X^{n-1}, X^{n-2}). \qquad \diamond$$

Lemma 4.2.14. $C_*^{cell}(X)$ *is a chain complex.*

Proof. We need only check that $\partial_{n-1}\partial_n = 0$. But this is the composition

$$H_n(X^n, X^{n-1}) \xrightarrow{\partial} H_{n-1}(X^{n-1}) \longrightarrow H_{n-1}(X^{n-1}, X^{n-2})$$

$$\xrightarrow{\partial} H_{n-2}(X^{n-2}) \longrightarrow H_{n-2}(X^{n-2}, X^{n-3})$$

and the middle two maps are two successive maps in the exact homotopy sequence of the pair (X^{n-1}, X^{n-2}), so their composition is the zero map. □

Definition 4.2.15. Let X be a CW-complex. Then the *cellular homology* of X is the homology of the cellular chain complex $C_*^{\text{cell}}(X)$ as defined in Definition A.2.2. ◇

Lemma 4.2.16. *The group $C_n^{\text{cell}}(X)$ is the free abelian group on the n-cells of X. If α_λ^n is the generator corresponding to the n-cell D_λ^n, $\lambda \in \Lambda_n$, then $\partial(\alpha_\lambda^n)$ is given as follows:*

$$H_n(D_\lambda^n, S_\lambda^{n-1}) \xrightarrow{\ \partial\ } H_{n-1}(S_\lambda^{n-1}) \xrightarrow{f_\lambda | S_\lambda^{n-1}} H_{n-1}(X^{n-1}) \longrightarrow H_{n-1}(X^{n-1}, X^{n-2}).$$

$$\cup\!\!\!| \qquad\qquad\qquad\qquad\qquad\qquad\qquad\qquad\qquad\qquad \cup\!\!\!|$$

$$\alpha_\lambda^n \longmapsto \hspace{7cm} \partial(\alpha_\lambda^n)$$

Proof. This follows directly from Lemma 4.2.11 and its proof. □

We let $Z_n^{\text{cell}}(X)$ denote $\text{Ker}(\partial_n) : C_n^{\text{cell}}(X) \to C_{n-1}^{\text{cell}}(X)$ and $B_n^{\text{cell}}(X)$ denote $\text{Im}(\partial_{n+1}) : C_{n+1}^{\text{cell}}(X) \to C_n^{\text{cell}}(X)$, so that

$$H_n^{\text{cell}}(X) = Z_n^{\text{cell}}(X)/B_n^{\text{cell}}(X).$$

Theorem 4.2.17. *Let X be a CW-complex. Suppose that X is finite dimensional or that H_* is compactly supported. Then the cellular homology of X is isomorphic to the ordinary homology of X.*

Proof. We begin with the following purely algebraic observation. Suppose we have three abelian groups and maps as shown:

$$H^1 \xleftarrow{\ k\ } H^2 \xrightarrow{\ j\ } H^3$$

where

1. k is a surjection.
2. j is an injection; set $Z^3 = \text{Im}(j)$.
3. There is a subgroup $B^3 \subseteq Z^3$ with $j^{-1}(B^3) = \text{Ker}(k)$.

Then $k \circ j^{-1} : Z^3/B^3 \to H^1$ is an isomorphism with inverse $j \circ k^{-1}$. To see this, note that $j : H^2 \to Z^3$ is an isomorphism, so $j^{-1} : Z^3 \to H^2$ is well-defined, and then we have isomorphisms

$$Z^3/B^3 \xrightarrow{\ j^{-1}\ } H^2/j^{-1}(B^3) = H^2/\text{Ker}(k) \cong H^1.$$

Note also that $j \circ k^{-1}$ is well-defined, as $j(\text{Ker}(k)) = B^3$.

We apply this here to construct isomorphisms, for each n,

$$\Theta_n : H_n^{\text{cell}}(X) \longrightarrow H_n(X).$$

Consider

$$H_n(X) \xleftarrow{k_n} H_n(X^n) \xrightarrow{j_n} H_n(X^n, X^{n-1})$$

with the maps induced by inclusion. We must verify the three conditions above.

We have already shown (1), that k_n is a surjection, in Corollary 4.2.12. Also, (2) is immediate from $0 = H_n(X^{n-1}) \to H_n(X^n) \to H_n(X^n, X^{n-1})$. Let us identify $Z^3 = \operatorname{Im}(j_n)$. We have

$$\partial_n : H_n(X^n, X^{n-1}) \xrightarrow{\partial} H_{n-1}(X^{n-1}) \xrightarrow{j_{n-1}} H_{n-1}(X^{n-1}, X^{n-2})$$

$$j_n \uparrow$$

$$H_n(X^n) \qquad\qquad 0 = H_{n-1}(X^{n-2})$$

Then j_{n-1} is an injection, so

$$\operatorname{Im}(j_n) = \operatorname{Ker}(\partial) = \operatorname{Ker}(j_{n-1} \circ \partial) = \operatorname{Ker}(\partial_n) = Z_n^{\text{cell}}(X).$$

As for (3), we have

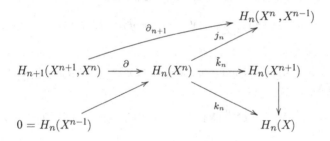

Note that $H_n(X^{n+1}) \to H_n(X)$ is an isomorphism. Also, $B_{n+1}^{\text{cell}}(X) = \operatorname{Im}(\partial_{n+1}) = \operatorname{Im}(j_n \circ \partial) \subseteq \operatorname{Im}(j_n)$. Then

$$j_n^{-1}\left(B_{n+1}^{\text{cell}}(X)\right) = j_n^{-1}\left(\operatorname{Im}(\partial_{n+1})\right) = j_n^{-1}\left(\operatorname{Im}(j_n\partial)\right)$$
$$= \operatorname{Im}(\partial) = \operatorname{Ker}(\tilde{k}_n) = \operatorname{Ker}(k_n).$$

Thus if $\Theta_n = k_n \circ j_n^{-1}$, we have an isomorphism

$$\Theta_n : Z_n^{\text{cell}}(X)/B_n^{\text{cell}}(X) = H_n^{\text{cell}}(X) \xrightarrow{\cong} H_n(X).$$

$\square$

Remark 4.2.18. Theorem 4.2.17 shows that the point of cellular homology is not that it is different from ordinary homology. Rather, for CW-complexes (where it is defined) it is the same. The point of cellular homology is that it is a better way of looking at homology. It is better for two reasons.

The first reason is psychological. It makes clear how the homology of a CW-complex comes from its cells.

The second reason is mathematical. If X is a finite complex, the cellular chain complex of X is finitely generated. $\Diamond$

This inherent finiteness not only makes cellular homology easier to work with, it allows us to effectively, and indeed easily, compute an important and very classical invariant of topological spaces as well.

Recall that any finitely generated abelian group A is isomorphic to $F \oplus T$, where F is a free abelian group of well-defined rank r (i.e., F is isomorphic to $\mathbb{Z}^r$) and T is a torsion group. In this case we define the *rank* of A to be r.

Definition 4.2.19. Let X be a space with $H_i(X)$ finitely generated for each i, and nonzero for only finitely many values of i. Then the *Euler characteristic* $\chi(X)$ is

$$\chi(X) = \sum_i (-1)^i \operatorname{rank} H_i(X).$$

$\Diamond$

Theorem 4.2.20. *Let X be a finite CW-complex. Then*

$$\chi(X) = \sum_i (-1)^i \cdot \text{number of } i\text{-cells of } X.$$

Proof. Let X have d_i i-cells and suppose $d_i = 0$ for $i > n$. We have the cellular chain complex of X

$$0 \longrightarrow C_n^{\text{cell}}(X) \longrightarrow C_{n-1}^{\text{cell}}(X) \longrightarrow \cdots \longrightarrow C_1^{\text{cell}}(X) \longrightarrow C_0^{\text{cell}}(X) \longrightarrow 0$$

with $C_i^{\text{cell}}(X)$ free abelian of rank d_i. But then it is purely algebraic result that

$$\sum_i (-1)^i d_i = \sum_i (-1)^i \operatorname{rank} H_i^{\text{cell}}(X)$$

and by Theorem A.2.13 this is equal to $\chi(X)$. $\square$

Remark 4.2.21. Note in particular that $\chi(X)$ is independent of the CW-structure on X. For example, let $X = S^n$. Then $\chi(X) = 2$ for n even and 0 for n odd. We have seen in Example 4.2.9 three different CW-structures on X. In the first, X has a single 0-cell and a single n-cell. In the second, X has two i-cells for each i between 0 and n. In the third, X has a single 0-cell, a single $(n-1)$-cell, and two n-cells. But counting cells in any of these CW-structures gives $\chi(X) = 2$ for n even and 0 for n odd. $\Diamond$

Remark 4.2.22. Let X be the surface of a convex polyhedron in $\mathbb{R}^3$. It is a famous theorem of Euler that, if V, E, and F denote the number of vertices, edges, and faces of X, then

$$V - E + F = 2.$$

But X is topologically S^2 and regarding X as the surface of a polyhedron gives a CW-structure on X with V 0-cells, E 1-cells, and F 2-cells, so this equation is a special case of Theorem 4.2.20.

For example, we may compute $V - E + F$ for each of the five Platonic solids.

Tetrahedron: $4 - 6 + 4 = 2$,
Cube: $8 - 12 + 6 = 2$,
Octahedron: $6 - 12 + 8 = 2$,
Dodecahedron: $20 - 30 + 12 = 2$,
Icosahedron: $12 - 30 + 20 = 2$. ◊

Recall that we considered covering spaces in Sect. 2.2. In general, if $\tilde{X}$ is a covering space X, there is no simple relationship between the homology groups of $\tilde{X}$ and the homology groups of X. There is, however, a simple relationship between their Euler characteristics.

Theorem 4.2.23. *Let X be a finite CW-complex. Let $\tilde{X}$ be an n-fold cover of X. Then $\chi(\tilde{X}) = n\chi(X)$.*

Proof. Given any cell decomposition of X, we may refine it to obtain a cell decomposition so that every cell is evenly covered by the covering projection. Then the inverse image of every cell is n cells, so the theorem immediately follows from Theorem 4.2.20. □

Example 4.2.24. Let R be a k-leafed rose, and let $\tilde{R}$ be any n-fold cover of R. Then R has one 0-cell and k 1-cells, so $\chi(R) = 1 - k$ (which of course agrees with $H_0(R) = \mathbb{Z}$ and $H_1(\mathbb{R}) = \mathbb{Z}^k$). Then $\chi(\tilde{R}) = n(1 - k)$. Now $H_0(\tilde{R}) = \mathbb{Z}$ (as by definition, a cover is connected), so we must have

$$1 - \text{rank}\, H_1(\tilde{R}) = n(1 - k)$$

and hence $H_1(\tilde{R}) = \mathbb{Z}^{n(k-1)+1}$. (Compare Corollary 2.4.6.) ◊

Definition 4.2.25. Let X and Y be CW-complexes. A *cellular map* $f : X \to Y$ is a map $f : X \to Y$ with the property that $f(X^n) \subseteq Y^n$ for every integer n. ◊

Lemma 4.2.26. *Let $f : X \to Y$ be a cellular map. Then for each i, f induces a map $f_i^{cell} : H_i^{cell}(X) \to H_i^{cell}(Y)$.*

Proof. By hypothesis, f induces a map $H_i(X^n, X^{n-1}) \to H_i(Y^n, Y^{n-1})$ for each i and n, and then it is easy to check this induces a map on cellular homology. □

Now given CW-complexes X and Y and a cellular map $f : X \to Y$, we have $f_i : H_i(X) \to H_i(Y)$ and $f_i^{cell} : H_i^{cell}(X) \to H_i^{cell}(Y)$. But we know that $H_i^{cell}(X)$ "agrees with" $H_i(X)$ and that $H_i^{cell}(Y)$ "agrees with" $H_i(Y)$. We would thus hope that f_i^{cell} "agrees with" f_i, and that is indeed the case.

Theorem 4.2.27. *Let X and Y be CW-complexes and let $f : X \to Y$ be a cellular map. Then the following diagram commutes:*

$$
\begin{array}{ccc}
H_i^{cell}(X) & \xrightarrow{\;\;\Theta_i\;\;} & H_i(X) \\
{\scriptstyle f_i^{cell}}\downarrow & & \downarrow{\scriptstyle f_i} \\
H_i^{cell}(Y) & \xrightarrow{\;\;\Theta_i\;\;} & H_i(Y)
\end{array}
$$

Proof. This follows easily from the commutativity of the diagram

$$
\begin{array}{ccccc}
H_i(X^{n-1}) & \longrightarrow & H_i(X^n) & \longrightarrow & H_i(X^n, X^{n-1}) \\
\downarrow & & \downarrow & & \downarrow \\
H_i(Y^{n-1}) & \longrightarrow & H_i(Y^n) & \longrightarrow & H_i(Y^n, Y^{n-1})
\end{array}
$$

where the vertical maps are all induced by f. $\square$

This theorem, together with the following theorem, known as the cellular approximation theorem, which we shall not prove, enables us to always use cellular homology to compute maps on homology between CW-complexes.

Theorem 4.2.28. *Let $f : X \to Y$ be an arbitrary map, where X and Y are CW-complexes. Then f is homotopic to a cellular map.*

Remark 4.2.29. All of this generalizes in a completely straightforward way to CW-pairs (X,A), where X is a CW-complex and A is a subcomplex, with chain groups

$$
C_n^{cell}(X,A) = H_n(X^n \cup A, X^{n-1} \cup A)
$$

and boundary maps

$$
H_n(X^n \cup A, X^{n-1} \cup A) \longrightarrow H_{n-1}(X^{n-1} \cup A) \longrightarrow H_{n-1}(X^{n-1} \cup A, X^{n-2} \cup A),
$$

and $C_n^{cell}(X,A)$ is isomorphic to the free abelian group on the n-cells of X that are not contained in A. $\Diamond$

Let us see that for cellular homology we have a strong form of excision.

Theorem 4.2.30. *Let (X,A) be a CW-pair and let $U \subseteq A$ be such that $(Y,B) = (X - U, A - U)$ is a CW-pair. Then the inclusion $(Y,B) \to (X,A)$ is excisive for cellular homology.*

Proof. First observe that the hypothesis on U implies that U is a union of open cells of A. Let F_n be the free abelian group on the n-cells of X that are not contained in A, which are exactly the n-cells of Y that are not contained in B. Then we have a commutative diagram

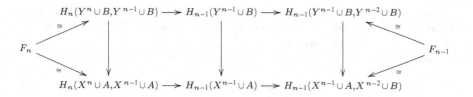

giving an isomorphism between the chain complexes $C_n^{\text{cell}}(Y,B)$ and $C_n^{\text{cell}}(X,A)$, and hence an isomorphism between their homology groups.   □

Recall that we defined the degree of a map $f : S^n \to S^n$ in Definition 4.1.8. With the use of cellular homology, we now show the following result. (Note this holds for any ordinary homology theory with $\mathbb{Z}$ coefficients as a consequence of the axioms. In the next chapter we will prove it for singular homology theory with $\mathbb{Z}$ coefficients by an entirely different geometric idea.)

Theorem 4.2.31. *Let d be any integer. Then for any integer $n \geq 1$, there exists a map of $f : S^n \to S^n$ of degree d.*

Proof. We prove this by induction on n. This is an argument where the base case $n = 1$ is the hardest, and the induction step is easy.

We claim that $f : S^1 \to S^1$ by $f(z) = z^d$ is a map of degree d, where we regard S^1 as the unit circle in the complex plane.

We now proceed to prove this claim.

Instead of working with S^1 directly it is easiest to first work with the interval $I = [0,1]$. We have $w : (I, \partial I) \to (S^1, 1)$ by $w(t) = \exp(2\pi i t)$ and we know w induces an isomorphism on homology $w_* : H_1(I, \partial I) \to H_1(S^1, 1) \cong H_1(S^1)$. (Indeed, w is a cellular map where I has the CW-structure with two 0-cells, the points of ∂I, and a single 1-cell c, and S^1 has the CW-structure with a single 0-cell 1 and a single 1-cell.)

Let c be the generator of $H_1(I, \partial I)$ with $\partial c = \{1\} - \{0\}$. We now give I a new CW-structure with 0-cells $p_k = k/d$, $k = 0,\dots,d$, and 1-cells I_k the subinterval $[p_{k-1}, p_k]$, $k = 1,\dots,d$. We let c_k be the generator of $H_1(I_k, \partial I_k)$ with $\partial c_k = p_k - p_{k-1}$. It is then easy to compute that $c_1 + \dots + c_d$ is the generator of $H_1(I, \partial I)$ with $\partial(c_1 + \dots + c_d) = \{1\} - \{0\}$, so $c_1 + \dots + c_d = c \in H_1(I, \partial I)$.

Let $\gamma = w_*(c)$ be a generator of S^1. (In fact $\gamma = \sigma_1$ is the standard generator.) Then

$$\gamma = w_*(c) = w_*(c_1 + \dots + c_d) = w_*(c_1) + \dots + w_*(c_d)$$

so

$$f_*(\gamma) = f_* w_*(c_1) + \dots + f_* w_*(c_d) = (fw)_*(c_1) + \dots + (fw)_*(c_d).$$

Now for $k = 1, \ldots, d$, let $v_k : I_k \to I$ by $v_k(t) = d(t - (k-1)/d)$. Then $v_k :$ $(I_k, \partial I_k) \to (I, \partial I)$ with $\partial((v_k)_*(c_k)) = \{1\} - \{0\} = \partial c$, so $(v_k)_*(c_k) = c$ (as $\partial :$ $H_1(I, \partial I) \to H_0(\partial I)$ is an injection).

But note that

$$fw = wv_k : I_k \longrightarrow S^1, \quad k = 1, \ldots, d.$$

Thus

$$
\begin{aligned}
f_*(\gamma) &= (fw)_*(c_1) + \cdots + (fw)_*(c_d) \\
&= (wv_1)_*(c_1) + \cdots + (wv_d)_*(c_d) \\
&= w_*((v_1)_*(c_1)) + \cdots + w_*((v_d)_*(c_d)) \\
&= w_*(c) + \cdots + w_*(c) = dw_*(c) = d\gamma
\end{aligned}
$$

as claimed, completing the proof of the $n = 1$ case.

The inductive step then follows immediately from Theorem 3.2.13. If $f : S^n \to S^n$ has degree d, then

$$\Sigma f : S^{n+1} \longrightarrow S^{n+1}$$

also has degree d (where Σ is the suspension).   □

Remark 4.2.32. This theory dualizes to obtain cellular cohomology $H^n_{\text{cell}}(X)$ for a CW-complex X (or $H^n_{\text{cell}}(X, A)$ for a CW-pair (X, A)). The cellular cochain complex of X is given by

$$C^n_{\text{cell}} = H^n(X^n, X^{n-1})$$

with $\delta^n : C^n_{\text{cell}}(X) \to C^{n+1}_{\text{cell}}(X)$ the composition

$$H^n(X^n, X^{n-1}) \longrightarrow H^n(X^n) \xrightarrow{\delta} H^{n+1}(X^{n+1}, X^n),$$

and similarly for (X, A).   ◇

Note that if X has only finitely many cells in each dimension, each cellular chain group, and hence each cellular cochain group, is a finitely generated free abelian group.

Theorem 4.2.33. *Let X be a CW-complex with only finitely many cells in each dimension. Then for each n, $H^{\text{cell}}_n(X)$ and $H^n_{\text{cell}}(X)$ are finitely generated abelian groups.*

Proof. $H^{\text{cell}}_n(X)$ is a quotient of $Z^{\text{cell}}_n(X)$, which is a subgroup of a finitely generated free abelian group, and hence itself is a finitely generated free abelian group, and similarly for $H^n_{\text{cell}}(X)$.   □

4.3   Real and Complex Projective Spaces

In this section we define real and complex projective spaces and compute their homology. As we will see, the computation is almost trivial for complex projective spaces but rather tricky for real projective spaces.

The first part of our development is almost identical in both cases so we handle them simultaneously. We let $\mathbb{F} = \mathbb{R}$ or $\mathbb{C}$.

Definition 4.3.1. The *projective space* $\mathbb{F}P^n$ is the space of lines through the origin in $\mathbb{F}^{n+1}$ with the following topology. Let S denote the unit sphere in $\mathbb{F}^{n+1}$. Then, for $\varepsilon > 0$, an ε-neighborhood of the line L_0 consists of all lines L whose intersection with S lies in an ε-neighborhood of $L_0 \cap S$. $\Diamond$

Note by lines here we mean $\mathbb{F}$-lines, i.e., $\mathbb{F}$-vector spaces of dimension 1.

Given a point $t = (t_1, \ldots, t_{n+1})$ of $\mathbb{F}^{n+1}$ other than the origin, we have the line L through the origin and t which we denote by $[t_1, \ldots, t_{n+1}]$; these are called the *homogeneous coordinates* of L. We thus have a map $\pi : \mathbb{F}^{n+1} - \{0\} \to \mathbb{F}P^n$ by $(t_1, \ldots, t_{n+1}) \mapsto [t_1, \ldots, t_{n+1}]$ with $\pi(tt_1, \ldots, tt_{n+1}) = \pi(t_1, \ldots, t_{n+1})$ for any $t \in \mathbb{F}^* = \mathbb{F} - 0$, and we may regard $\mathbb{F}P^n$ as the quotient of the space $\mathbb{F}^{n+1} - 0$ under this action of $\mathbb{F}^*$. Alternatively, restricting our attention to points $(t_1, \ldots, t_{n+1})$ of (Euclidean) norm 1, i.e., the unit sphere S, we have an action of $G = \{t \in \mathbb{P}^* \mid |t| = 1\}$ on S and $\mathbb{F}P^n$ is the quotient of S by this action. (It is routine to check that the topology of $\mathbb{F}P^n$ agrees with the quotient topology from these actions.) If $\mathbb{F} = \mathbb{C}$, then $S = S^{2n+1}$ and $G = \{z \in \mathbb{C} \mid |z| = 1\}$ is the unit circle. If $\mathbb{F} = \mathbb{R}$, then $S = S^n$ and $G = \{x \in \mathbb{R} \mid |x| = 1\} = \{\pm 1\}$.

We may regard $\mathbb{F}^n \subset \mathbb{F}^{n+1}$ as the subspace of points with last coordinate $t_{n+1} = 0$. Then we get a corresponding inclusion $\mathbb{F}P^{n-1} \subset \mathbb{F}P^n$ with $\mathbb{F}P^{n-1} = \{[t_1, \ldots, t_n, 0]\} \subset \mathbb{F}P^n$.

Theorem 4.3.2. *Let $d = \dim_{\mathbb{R}} \mathbb{F}$ (so that $d = 1$ if $\mathbb{F} = \mathbb{R}$ and $d = 2$ if $\mathbb{F} = \mathbb{C}$). Then $\mathbb{F}P^n$ has a CW-structure with one cell in dimension di for each $i = 0, \ldots, n$.*

Proof. By induction on n.

For $n = 0$, $\mathbb{F}P^0$ is just a point.

Assume now the theorem is true for $n - 1$. We shall show that $\mathbb{F}P^n - \mathbb{F}P^{n-1}$ is a single cell of dimension dn, which, by induction, completes the proof.

Now $\mathbb{F}P^n - \mathbb{F}P^{n-1} = \{[t_1, \ldots, t_{n+1}] \mid t_{n+1} \neq 0\}$. Let D be the unit ball in $\mathbb{F}^n$, $D = \{(t_1, \ldots, t_n) \mid \sum |t_i|^2 \leq 1\}$, and let S be its boundary.

We have a map $(D, S) \to (\mathbb{F}P^n, \mathbb{F}P^{n-1})$ given by

$$(t_1, \ldots, t_n) \mapsto \left[t_1, \ldots, t_n, \sqrt{1 - |t|^2} \right],$$

where $|t|^2 = t_1^2 + \cdots + t_n^2$, and where we choose the positive square root.

Observe that this map restricts to a homeomorphism from $\mathrm{int}(D)$ to $\mathbb{F}P^n - \mathbb{F}P^{n-1}$, (and on S it is the map $(t_1, \ldots, t_n) \mapsto [t_1, \ldots, t_n, 0]$ so S maps onto $\mathbb{F}P^{n-1}$).

Thus $\mathbb{F}P^n$ is obtained from $\mathbb{F}P^{n-1}$ by adjoining D. But D is a single cell of (real) dimension dn.   □

Theorem 4.3.3. *The homology of* $\mathbb{C}P^n$ *is as follows:*

$$H_i(\mathbb{C}P^n) = \begin{cases} 0 & i > 2n \\ \mathbb{Z} & 0 \leq i \leq 2n \text{ even} \\ 0 & 0 < i < 2n \text{ odd.} \end{cases}$$

Proof. The cellular chain complex of $\mathbb{C}P^n$ is

$$0 \longrightarrow \mathbb{Z} \longrightarrow 0 \longrightarrow \mathbb{Z} \longrightarrow \cdots \longrightarrow \mathbb{Z} \longrightarrow 0 \longrightarrow \mathbb{Z} \longrightarrow 0$$

with $\mathbb{Z}$ in every even dimension between 0 and $2n$, and 0 otherwise.   □

Theorem 4.3.4. *The homology of* $\mathbb{R}P^n$ *is as follows:*

$$H_i(\mathbb{R}P^n) = \begin{cases} 0 & i > n \\ 0 & i = n, \ n \text{ even} \\ \mathbb{Z} & i = n, \ n \text{ odd} \\ \mathbb{Z}_2 & 0 < i < n, \ i \text{ odd} \\ 0 & 0 < i < n, \ i \text{ even} \\ \mathbb{Z} & i = 0. \end{cases}$$

Proof. The cellular chain complex of $\mathbb{R}P^n$ is, by Lemma 4.2.11,

$$0 \longrightarrow \mathbb{Z} \xrightarrow{\partial_n} \mathbb{Z} \xrightarrow{\partial_{n-1}} \mathbb{Z} \longrightarrow \cdots \mathbb{Z} \xrightarrow{\partial_1} \mathbb{Z} \longrightarrow 0$$

with $\mathbb{Z}$ occurring in every dimension between 0 and n. We shall show that (with proper choice of generator), ∂_i is multiplication by $1 + (-1)^i$, which yields the result.

Again we proceed by induction on n. For $n = 0$, $\mathbb{R}P^0$ is a point, and its homology is as stated. For $n = 1$, we have the chain complex $0 \to \mathbb{Z} \xrightarrow{\partial_1} \mathbb{Z} \to 0$ and since $H_0(\mathbb{R}P^1) = \mathbb{Z}$, we must have ∂_1 the 0 map. (Actually, we first encountered $\mathbb{R}P^n$ in Example 2.3.4, where we observed that $\mathbb{R}P^1 = S^1$, and we know the homology of S^1.)

We know that the inclusion of $\mathbb{R}P^{n-1}$ into $\mathbb{R}P^n$ induces isomorphisms on homology except possibly in dimensions $n-1$ and n, by Lemma 4.2.11.

By induction, we only need to determine the map

$$\mathbb{Z} \xrightarrow{\partial_n} \mathbb{Z}$$

in the above chain complex.

There are two cases, n even and n odd.

First consider the case n odd. Then the above map is

$$H_n(\mathbb{R}P^n, \mathbb{R}P^{n-1}) \xrightarrow{\partial_n} H_{n-1}(\mathbb{R}P^{n-1}, \mathbb{R}P^{n-2}).$$

But recall that this map was defined to be the composition

$$H_n(\mathbb{R}P^n, \mathbb{R}P^{n-1}) \longrightarrow H_{n-1}(\mathbb{R}P^{n-1}) \longrightarrow (\mathbb{R}P^{n-1}, \mathbb{R}P^{n-2})$$

and by the inductive hypothesis $H_{n-1}(\mathbb{R}P^{n-1}) = 0$ (as $n-1$ is even) so this must be the 0 map.

Now consider the case n even.

Let $\pi : S^n \to \mathbb{R}P^n$ be the 2-fold covering map, i.e., $\pi((x_1, \ldots, x_{n+1})) = [x_1, \ldots, x_{n+1}]$.

Let $T^+ : (D^n, S^{n-1}) \to (S^n, S^{n-1})$ be the map $T(x_1, \ldots, x_n) = (x_1, \ldots, x_n, \sqrt{1 - |x|^2})$, where $|x|^2 = x_1^2 + \cdots + x_n^2$, and let $T^- : (D^n, S^{n-1}) \to (S^n, S^{n-1})$ be the composition $T^- = a \circ T^+$, where $a : D^n \to D^n$ is the antipodal map, $a(x_1, \ldots, x_n) = (-x_1, \ldots, -x_n)$. Note that T^+ and T^- are the attaching maps of two n-cells to S^{n-1} in forming a CW-structure on S^n. Note also that if $f : (D^n, S^{n-1}) \to (\mathbb{R}P^n, \mathbb{R}P^{n-1})$ is the attaching map of the n-cell to $\mathbb{R}P^{n-1}$ in forming a CW-structure on $\mathbb{R}P^n$, then $f = \pi \circ T^+ = \pi \circ T^{-1}$. Since $f_* : H_n(D^n, S^{n-1}) \to H_n(\mathbb{R}P^n, \mathbb{R}P^{n-1})$ is an isomorphism, it follows that $\pi_* : H_n(S^n, S^{n-1}) \to H_n(\mathbb{R}P^n, \mathbb{R}P^{n-1})$ is an epimorphism.

Now consider the commutative diagram

$$
\begin{array}{ccccccccc}
0 & \longrightarrow & H_{n-1}(S^{n-1}) & \longrightarrow & H_{n-1}(S^{n-1}, S^{n-2}) & \longrightarrow & H_{n-2}(S^{n-2}) & \longrightarrow & 0 \\
 & & \big\downarrow a_* & & \big\downarrow a_* & & \big\downarrow a_* & & \\
0 & \longrightarrow & H_{n-1}(S^{n-1}) & \longrightarrow & H_{n-1}(S^{n-1}, S^{n-2}) & \longrightarrow & H_{n-2}(S^{n-2}) & \longrightarrow & 0
\end{array}
$$

Let e denote a generator of $H_{n-1}(D^{n-1}, S^{n-2}) \cong \mathbb{Z}$. Then $H_{n-1}(S^{n-1}, S^{n-2}) \cong \mathbb{Z} \oplus \mathbb{Z}$ generated by $t^+ = T_*^+(e)$ and $t^- = T_*^-(e)$. We have seen in Lemma 4.1.9 that $a_* : H_i(S^i) \to H_i(S^i)$ has degree $(-1)^{i+1}$, so $a_* : H_{n-2}(S^{n-2}) \to H_{n-2}(S^{n-2})$ is multiplication by -1. Also, $a_*(t^+) = t^-$ and $a_*(t^-) = t^+$. Chasing the right-hand square, this shows that $t^+ + t^-$ is in the kernel of $H_{n-1}(S^{n-1}, S^{n-2}) \to H_{n-2}(S^{n-2})$, and, since t^+ cannot be in the kernel (as then this map would be the zero map, while by exactness it must be an epimorphism), $t^+ + t^-$ in fact generates the kernel. Thus, by exactness, there is a generator g of $H_{n-1}(S^{n-1})$ whose image in $H_{n-1}(S^{n-1}, S^{n-2})$ is $t^+ + t^-$.

Now we have a commutative diagram

$$\begin{array}{ccccc}
H_n(S^n,S^{n-1}) & \longrightarrow & H_{n-1}(S^{n-1}) & \xrightarrow{\;i_*\;} & H_{n-1}(S^{n-1},S^{n-2}) \\
\ \downarrow{\scriptstyle\pi_*} & & \ \downarrow{\scriptstyle\pi_*} & \ \searrow{\scriptstyle\rho_*} & \ \downarrow{\scriptstyle\pi_*} \\
H_n(\mathbb{R}P^n,\mathbb{R}P^{n-1}) & \xrightarrow{\;\partial\;} & H_{n-1}(\mathbb{R}P^{n-1}) & \xrightarrow{\;j_*\;} & H_{n-1}(\mathbb{R}P^{n-1},\mathbb{R}P^{n-2})
\end{array}$$

where the composition of the two maps in the bottom row is the boundary map in the cellular chain complex of $\mathbb{R}P^n$. We have observed that the left-hand vertical map is an epimorphism, and, by the exact sequence of the pair (S^n,S^{n-1}), the left-hand horizontal map in the top row is an epimorphism as well. Thus the image of $\partial : H_n(\mathbb{R}P^n,\mathbb{R}P^{n-1}) \to H_{n-1}(\mathbb{R}P^{n-1})$ agrees with the image of $\pi_* : H_{n-1}(S^{n-1}) \to H_{n-1}(\mathbb{R}P^{n-1})$. Hence there is a generator e of $H_n(\mathbb{R}P^n,\mathbb{R}P^{n-1})$ with $\partial(e) = \pi_*(g)$.

To compute the boundary map in the cellular chain complex we must find $j_*(\partial(e))$. By commutativity that is equal to $\rho_*(g)$.

Now consider the right-hand square in the diagram. By construction, $i_*(g) = t^+ + t^-$. (Also, j_* is an isomorphism, by the exact sequence of the pair $(\mathbb{R}P^{n-1},\mathbb{R}P^{n-2})$.) Observe that $\pi \circ a = \pi$. Since $H_{n-1}(S^{n-1},S^{n-2}) = \mathbb{Z} \oplus \mathbb{Z}$ is generated by the two classes t^+ and t^-, and $t^- = a_*(t^+)$, and $\pi_* : H_{n-1}(S^{n-1}) \to H_{n-1}(\mathbb{R}P^{n-1})$ is an epimorphism, we see that $H_{n-1}(\mathbb{R}P^{n-1},\mathbb{R}P^{n-2})$ is generated by $u = \pi_*(t^+)$.

But then

$$\begin{aligned}
j_*(\partial(e)) = \rho_*(g) &= \pi_*(i_*(g)) \\
&= \pi_*(t^+ + t^-) \\
&= \pi_*(t^+) + \pi_*(t^-) \\
&= \pi_*(t^+) + \pi_*(a_*(t^+)) \\
&= \pi_*(t^+) + (\pi a)_*(t^+) \\
&= \pi_*(t^+) + \pi_*(t^+) \\
&= 2u.
\end{aligned}$$

Thus, with this choice of generators,

$$\mathbb{Z} \xrightarrow{\;\partial_n\;} \mathbb{Z}$$

is a multiplication by 2, as claimed, and we are done. $\square$

4.4 Exercises

Exercise 4.4.1. Complete the proof of Lemma 4.2.11.

Exercise 4.4.2. Verify the proof of Lemma 4.2.16.

Exercise 4.4.3. Complete the proof of Lemma 4.2.26.

Exercise 4.4.4. Let $f : S^n \to S^n$ be fixed-point free, i.e., $f(x) \neq x$ for every $x \in S^n$. Show that f has degree $(-1)^{n+1}$.

Exercise 4.4.5. Let $f : S^n \to S^n$ be a map with nonzero degree. Show that f is onto S^n, i.e., that for every $y \in S^n$ there is an $x \in S^n$ with $f(x) = y$.

Exercise 4.4.6. Let S be the Riemann sphere, i.e., $S = \mathbb{C} \cup \{\infty\}$ where $\mathbb{C}$, the complex plane, has its usual topology, and a basis for the neighborhoods of the point ∞ is $\{\infty\} \cup \{z \in \mathbb{C} \mid |z| > N\}$ for $N = 1, 2, 3, \ldots$. Show that S is homeomorphic to S^2.

Exercise 4.4.7. Let p be a polynomial with complex coefficients, which we regard as defining a map $p : \mathbb{C} \to \mathbb{C}$ by $z \mapsto p(z)$. Show that p extends to a map $\tilde{p} : S \to S$ as follows: If $p(z) = a_0$ is a constant polynomial, then $\tilde{p}(\infty) = a_0$. If p is a nonconstant polynomial, then $\tilde{p}(\infty) = \infty$. (Of course, map means continuous map.)

Exercise 4.4.8. Suppose that p is a polynomial of degree d. (Here we define the degree of any constant polynomial to be 0.) Show that $\tilde{p} : S \to S$ is a map of degree d.

Exercise 4.4.9. Prove the Fundamental Theorem of Algebra: Every nonconstant complex polynomial has a complex root.

Exercise 4.4.10. Let $r = p/q$ be a quotient of nonzero complex polynomials. Show that r naturally defines a map $\tilde{r} : S \to S$. Furthermore, if p and q are polynomials of degree m and n respectively, show that $\tilde{r}$ is a map of degree $m - n$.

Exercise 4.4.11. Compute the homology of the 2-torus T by a Mayer-Vietoris argument. T is

Exercise 4.4.12. Compute the homology of S_g, the surface of genus g. These are given by $S_0 = S^2$, the 2-sphere; $S_1 = T$, the 2-torus, and S_g for $g > 1$ as pictured below:

S_2 S_3 etc.

Exercise 4.4.13. Let $X = X_1 \cup X_2$, $A = X_1 \cap X_2$, and suppose that the inclusion $(X_1, A) \to (X, X_2)$ is excisive. If $H_*(A)$, $H_*(X_1)$, and $H_*(X_2)$ are all finitely generated, show that $H_*(X)$ is finitely generated and furthermore that

$$\chi(X) = \chi(X_1) + \chi(X_2) - \chi(A).$$

Exercise 4.4.14. A Platonic solid is a tessellation of the surface of a convex polyhedron in $\mathbb{R}^3$ by mutually congruent polygons. Show that the only Platonic solids are the five listed in Remark 4.2.22.

Exercise 4.4.15. Let V and F be any positive integers and let E be any nonnegative integer with $V - E + F = 2$. Show there is a CW structure on S^2 with V 0-cells, E 1-cells, and F 2-cells.

Exercise 4.4.16. Suppose X and Y are CW-complexes. Show that $X \times Y$ is a CW-complex.

Exercise 4.4.17. Suppose X and Y are finite CW-complexes. Show that $\chi(X \times Y) = \chi(X)\chi(Y)$.

Exercise 4.4.18. Let X be a CW-complex with $X = X_1 \cup X_2$ with X_1, X_2, and $A = X_1 \cap X_2$ subcomplexes. Show that if any three of $H_*(X)$, $H_*(X_1)$, $H_*(X_2)$, and $H_*(A)$ are finitely generated, so is the fourth. (Observe in this case $\chi(X) = \chi(X_1) + \chi(X_2) - \chi(A)$.)

Exercise 4.4.19. Give an example of a CW complex X with $H_*(X)$ finitely generated, but $X = X_1 \cup X_2$ with X_1, X_2, and $A = X_1 \cap X_2$ subcomplexes with none of $H_*(X_1)$, $H_*(X_2)$, and $H_*(A)$ finitely generated.

Exercise 4.4.20. Recall that X is an m-dimensional CW-complex if it has at least one m-cell, but no k-cells for $k > m$. Suppose that X is an m-dimensional CW-complex and that Y is an n-dimensional CW-complex with $m \neq n$. Show that X and Y are not homeomorphic.

Chapter 5
Singular Homology Theory

In this chapter we develop singular homology, an ordinary homology theory, and derive many of its properties.

We begin with $H_n(X)$ or $H_n(X;\mathbb{Z})$ (resp. $H_n(X,A)$ or $H_n(X,A;\mathbb{Z})$), which is an ordinary homology theory with $\mathbb{Z}$ coefficients. But we further derive from this singular homology theory with arbitrary coefficients, as well as singular cohomology theory (with arbitrary coefficients).

5.1 Development of the Theory

In this section we construct a nontrivial homology theory, singular homology. It is an ordinary homology theory, and the coefficient group is the integers $\mathbb{Z}$.

There are two approaches to singular homology: via singular cubes or singular simplices. Each has advantages and disadvantages, but for our purposes singular cubes are much preferable, and so we use them. (Both approaches yield the same results. The question is which is technically simpler.) In particular, the use of singular cubes makes it easier to derive chain homotopies from homotopic maps, and makes it easier to derive results on the homology of product spaces.

We begin with the standard cube and its faces.

Definition 5.1.1. The *standard* 0-*cube* I^0 is the point $0 \in \mathbb{R}^0$. For $n \geq 1$, the *standard n-cube* is

$$I^n = \{x = (x_1, x_2, \ldots, x_n) \in \mathbb{R}^n \mid 0 \leq x_i \leq 1, \ i = 1, \ldots, n\}.$$

Its i-th *front face* is

$$A_i = A_i(I^n) = \{x \in I^n \mid x_i = 0\}$$

© Springer International Publishing Switzerland 2014
S.H. Weintraub, *Fundamentals of Algebraic Topology*, Graduate Texts
in Mathematics 270, DOI 10.1007/978-1-4939-1844-7_5

and its i-th *back face* is

$$B_i = B_i(I^n) = \{x \in I^n \mid x_i = 1\}.$$
◊

Definition 5.1.2. Let I^n be the standard n-cube. Its *boundary* is given by

$$\partial I^0 = 0$$

and

$$\partial I^0 = \sum_{i=1}^{n} (-1)^i (A_i - B_i)$$

for $n > 0$.
◊

In this definition, ∂I^n is considered to be an element in the free abelian group generated by $\{A_i, B_i \mid i = 1, \dots, n\}$. We then have the following basic lemma:

Lemma 5.1.3. *For any n, $\partial(\partial I^n) = 0$.*

Proof. For $n \le 1$ this is clear.

For $n \ge 2$, $\partial(\partial I^n)$ is an element in the free abelian group generated by the $(n-2)$-faces of I^n, i.e., by the subsets, for each $i \ne j$ and each $\varepsilon_i = 0$ or 1, $\varepsilon_j = 0$ or 1,

$$\{x \in I^n \mid x_i = \varepsilon_i, \ x_j = \varepsilon_j\}.$$

Geometrically, each $(n-2)$-face of I^n is a free of two $(n-1)$-faces, and the signs are chosen in Definition 5.1.2 so that they cancel. This is a routine but tedious calculation.
□

Definition 5.1.4. Let X be a topological space.

A *singular n-cube* of X is a map $\Phi : I^n \to X$.
◊

We let $\alpha_i : I^{n-1} \to I^n$ be the inclusion of the i-th front face and $\beta_i : I^{n-1} \to I^n$ be the inclusion of the i-th back face, i.e.

$$\alpha_i(x_1, \dots, x_{n-1}) = (x_1, \dots, x_{i-1}, 0, x_i, \dots, x_{n-1})$$
$$\beta_i(x_1, \dots, x_{n-1}) = (x_1, \dots, x_{i-1}, 1, x_i, \dots, x_{n-1}).$$

Definition 5.1.5. For $n \ge 2$, a singular n-cube $f : I^n \to X$ is degenerate if the value of f is independent of at least one coordinate x_i, i.e., if there is a singular $(n-1)$-cube $\Psi : I^{n-1} \to X$ with

$$\Phi(x_1, \dots, x_n) = \Psi(x_1, \dots, x_{i-1}, x_{i+1}, \dots, x_n).$$

A singular 0-cube is always non-degenerate.
◊

Observe that in the degenerate case we have, in particular,

$$\Phi\alpha_i = \Phi\beta_i = \Psi : I^{n-1} \longrightarrow X, \quad \text{for } n \geq 1.$$

Definition 5.1.6. Let X be a topological space. The group $Q_n(X)$ is the free abelian group generated by the singular n-cubes of X. The subgroup $D_n(X)$ is the free abelian group generated by the degenerate singular n-cubes. (In particular, $D_0(X) = \{0\}$.)

For each $n \geq 0$, the boundary map $\partial_n^Q = \partial^Q$ is defined by $\partial^Q = 0$ if $n = 0$, and if $n \geq 1$, and $\Phi : I^n \to X$ is a singular n-cube,

$$\partial^Q \Phi = \sum_{i=1}^{n} (-1)^i (A_i\Phi - B_i\Phi) \in Q_{n-1}(X)$$

where $A_i\Phi = \Phi\alpha_i : I^{n-1} \to X$ and $B_i\Phi = \Phi\beta_i : I^{n-1} \to X$. ◇

Lemma 5.1.7. (1) $\partial_{n-1}^Q \partial_n^Q : Q_n(X) \to Q_{n-2}(X)$ *is the 0 map.*
(2) $\partial^Q(D_n(X)) \subseteq D_{n-1}(X)$.

Corollary 5.1.8. *Let* $C_n(X) = Q_n(X)/D_n(X)$. *Then* $\partial_n^Q : Q_n(X) \to Q_{n-1}(X)$ *induces* $\partial_n : C_n(X) \to C_{n-1}(X)$ *with* $\partial_{n-1}\partial_n = 0$.

Definition 5.1.9. The chain complex $C(X)$:

$$\cdots \longrightarrow C_2(X) \xrightarrow{\partial_2} C_1(X) \xrightarrow{\partial_1} C_0(X) \xrightarrow{\partial_0} 0 \longrightarrow 0 \longrightarrow \cdots$$

is the *singular chain complex* of X. ◇

Definition 5.1.10. The homology of the singular chain complex of X is the *singular homology* of X. ◇

To quote the definition of the homology of a chain complex from Sect. A.2:

$Z_n(X) = \text{Ker}\,(\partial_n : C_n(X) \longrightarrow C_{n-1}(X))$, the group of *singular n-cycles*,

$B_n(X) = \text{Im}\,(\partial_{n+1} : C_{n+1}(X) \longrightarrow C_n(X))$, the group of *singular n-boundaries*,

$H_n(X) = Z_n(X)/B_n(X)$, the n-th *singular homology group* of X

We now define the homology of a pair.

Definition 5.1.11. Let (X,A) be a pair. The relative singular chain complex $C(X,A)$ is the chain complex

$$\cdots \longrightarrow C_2(X)/C_2(A) \xrightarrow{\partial} C_1(X)/C_1(A) \xrightarrow{\partial} C_0(X)/C_0(A) \xrightarrow{\partial} 0 \longrightarrow 0 \longrightarrow \cdots.$$

Its homology is the *singular homology* of the pair (X,A). ◇

Finally, we define the induced map on homology of a map of spaces, or of pairs.

Lemma 5.1.12. *Let $f : X \to Y$ be a map. Then f induces a chain map $\{f_n : C_n(X) \to C_n(Y)\}$ where $f_n : C_n(X) \to C_n(Y)$ as follows. Let $\Phi : I^n \to X$ be a singular n-cube. Then $f_n\Phi = f\Phi : I_n \to Y$ where $f\Phi$ denotes the composition. Similarly $f : (X,A) \to (Y,B)$ induces a map $f_n : C_n(X,A) \to C_n(Y,B)$ by composition.*

This chain map induces a map on singular homology $\{f_n : H_n(X) \to H_n(Y)\}$ and similarly $\{f_n : H_n(X,A) \to H_n(Y,B)\}$.

Proof. This would be immediate if we were dealing with $Q_n(X)$ and $Q_n(Y)$. But since $f\Phi$ is degenerate wherever Φ is, it is just about immediate for $C_n(X)$ and $C_n(Y)$.

Then the fact that we have maps on homology is a direct consequence of Lemma A.2.7.   □

Definition 5.1.13. The above maps $\{f_n : H_n(X) \to H_n(Y)\}$ or $\{f_n : H_n(X,A) \to H_n(Y,B)\}$ are the *induced* maps on singular homology by the map $f : X \to Y$ or the map $f : (X,A) \to (Y,B)$.   ◊

We now verify that singular homology satisfies the Eilenberg-Steenrod axioms.

Theorem 5.1.14. *Singular homology satisfies Axioms 1 and 2.*

Proof. Immediate from the definition of the induced map on singular cubes as composition.   □

Theorem 5.1.15. *Singular homology satisfies Axiom 3.*

Proof. Immediate from the definition of the boundary map on singular cubes and from the definition of the induced map on singular cubes as composition.   □

Theorem 5.1.16. *Singular homology satisfies Axiom 4.*

Proof. We have defined $C_n(X,A) = C_n(X)/C_n(A)$. Thus for every n, we have a short exact sequence

$$0 \longrightarrow C_n(A) \longrightarrow C_n(X) \longrightarrow C_n(X,A) \longrightarrow 0.$$

In other words, we have a short exact sequence of chain complexes

$$0 \longrightarrow C_*(A) \longrightarrow C_*(X) \longrightarrow C_*(X,A) \longrightarrow 0.$$

But then we have a long exact sequence in homology by Theorem A.2.10.   □

Theorem 5.1.17. *Singular homology satisfies Axiom 5.*

Proof. For simplicity we consider the case of homotopic maps of spaces $f : X \to Y$ and $g : X \to Y$ (rather than maps of pairs). Then by definition, setting $f_0 = f$ and $f_1 = g$, there is a map $F : X \times I \to Y$ with $F(x,0) = f_0(x)$ and $F(x,1) = f_1(x)$.

Define a map $\tilde{F} : C_n(X) \to C_{n+1}(Y)$ as follows. Let $\Phi : I^n \to X$ be a singular n-cube. Then $\tilde{F}\Phi : I^{n+1} \to Y$ is defined by

$$\tilde{F}\Phi(x_1,\dots,x_{n+1}) = F(\Phi(x_1,\dots,x_n),x_{n+1}).$$

Then it is routine (but lengthy) to check that F provides a chain homotopy between $f_* : C_*(X) \to C_*(Y)$ and $g_* : C_*(X) \to C_*(Y)$, so that $f_* = g_* : H_*(X) \to H_*(Y)$ by Lemma A.2.9.   □

Theorem 5.1.18. *Singular homology satisfies Axiom 6.*

We shall not prove that singular homology satisfies Axiom 6, the excision axiom. This proof is quite involved. We merely give a quick sketch of the basic idea. Consider a singular chain, say c, the singular 1-cube which is the illustrated path

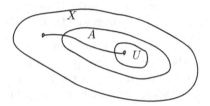

This chain is too large, and we subdivide it as illustrated.

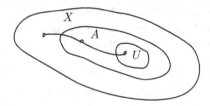

Then we show c is homologous to $c_1 + c_2$ with c_1 (resp. c_2) the left (resp. right) subpaths in $C_*(X,A)$. But c_1 is in the image of the inclusion $C_*(X-U,A-U) \to C_*(X,A)$.

Let us now compute the singular homology of a point.

Theorem 5.1.19. *Let X be the space consisting of a single point. Then $H_0(X) \cong \mathbb{Z}$ and $H_i(X) = 0$ for $i \neq 0$. Thus singular homology satisfies the dimension axiom, Axiom 7, and has coefficient group $\mathbb{Z}$.*

Proof. Let $\Phi : I^0 \to X$ be the unique map. Then $C_0(X)$ is the free abelian group generated by Φ. On the other hand, for any $i > 0$, $\Phi : I^i \to X$ is a degenerate i-cube. Hence $C_i(X) = \{0\}$ for $i > 0$. Thus $C_*(X)$ is the chain complex

$$\cdots \longrightarrow 0 \longrightarrow 0 \longrightarrow \mathbb{Z} \longrightarrow 0 \longrightarrow 0 \longrightarrow \cdots$$

with homology as stated.   □

Remark 5.1.20. It is to make Theorem 5.1.19 hold that we must use $C_i(X) = Q_i(X)/D_i(X)$ rather than work with $Q_i(X)$ itself. Let X be a point. Then in $Q_i(X)$ we have, for each i, the unique map $\Phi : I^i \to X$, and $\partial_i^Q \Phi = 0$ as the front and back faces cancel each other out. Then $Q_*(X)$ is the chain complex

$$\cdots \longrightarrow \mathbb{Z} \longrightarrow \mathbb{Z} \longrightarrow \mathbb{Z} \longrightarrow 0 \longrightarrow 0 \longrightarrow \cdots$$

with each boundary map being the 0 map, and this chain complex has homology $\mathbb{Z}$ in every nonnegative dimension.

However, it is not just that this complex gives the "wrong" answer. Rather, it is that it gives the wrong answer for a stupid reason, the presence of all these geometrically meaningless singular cubes. So we divide out by them to get a geometrically meaningful theory.   ◇

As in Chap. 4, we once and for all establish an isomorphism between $H_0(*)$ and $\mathbb{Z}$ by choosing a generator $1_* \in H_0(*)$ which we identify with $1 \in \mathbb{Z}$.

Definition 5.1.21. The class $1_* \in H_0(*)$ is the homology class represented by the unique map $\Phi : I^0 \to *$. Then for any space p consisting of a simple point, the class $1_p \in H_0(p)$ is $1_p = f_0(1_*)$ where $f : * \to p$ is the unique map. (More simply, $I_p \in H_0(p)$ is the homology class represented by the unique map $\Phi : I^0 \to p$.)   ◇

Let us make a definition and a pair of observations.

Definition 5.1.22. Let X be a space and let $c \in C_n(X)$ be a singular chain. Choose a representative of c of the form $\sum_{i=0}^N m_i \Phi_i$ where $m_i \neq 0$ for each i and Φ_i is a non-degenerate singular n-cube for each i. The *support* of c, supp(c), is defined to be $\emptyset$ if $N = 0$, and otherwise supp$(c) = \bigcup_{i=0}^N \Phi_i(I^n) \subseteq X$.   ◇

Lemma 5.1.23. *For any singular chain c,* supp$(\partial c) \subseteq$ supp(c).

Theorem 5.1.24. *For any singular chain c,* supp(c) *is a compact subset of X.*

Proof. For any $\Phi : I^n \to X$, $\Phi(I^n)$ is a compact subset of X as it is the continuous image of a compact set. Then for any singular chain c, supp(c) is a finite union of compact sets and hence is compact.   □

Corollary 5.1.25. *Let X be a union of components, $X = \bigcup_{i \in I} X_i$. Then for any n, $H_n(X) = \bigoplus_{i \in I} H_n(X_i)$.*

Proof. This follows for any generalized homology theory from Lemma 3.2.1 if there are only finitely many components. But for singular homology theory, if $c \in C_n(X)$ is any chain, then supp(c) is compact, by Theorem 5.1.24, so is contained in $\bigcup_{i \in J} X_i$ for some finite subset J of I. Thus $C_n(X) = \bigoplus_{i \in I} C_n(X_i)$. But clearly $\partial : C_n(X_i) \to C_{n-1}(X_{i-1})$ for any i, so as chain complexes $C_*(X) = \bigoplus_{i \in I} C_*(X_i)$ and hence $H_*(X) = \bigoplus_{i \in I} H_*(X_i)$.   □

We record the following result for future use.

Lemma 5.1.26. (1) *For any space X, the group of singular n-chains $C_n(X)$ is isomorphic to the free abelian group with basis the non-degenerate n-cubes.*

(2) *For any pair (X,A), the group $C_n(X,A)$ is isomorphic to the free abelian group with basis those non-degenerate n-cubes whose support is not contained in A.*

(3) *The short exact sequence*

$$0 \longrightarrow C_n(A) \longrightarrow C_n(X) \longrightarrow C_n(X,A) \longrightarrow 0$$

splits, and hence $C_n(X)$ is isomorphic to $C_n(A) \oplus C_n(X,A)$.

Proof. This follows easily once we recall that $C_n(X) = Q_n(X)/D_n(X)$ where $Q_n(X)$ is the free abelian group on all n-cubes and $D_n(X)$ is the free abelian group on the degenerate n-cubes   □

We also record the following definition.

Definition 5.1.27. A space X is of *finite type* if $H_n(X)$ is finitely generated for each n.   ◇

5.2 The Geometric Meaning of H_0 and H_1

In this section we see the geometric content of the singular homology groups $H_0(X)$ and $H_1(X)$.

Theorem 5.2.1. *Let X be a space. Then $H_0(X)$ is isomorphic to the free abelian group on the path components of X.*

Proof. We assume X nonempty. We have already seen in Corollary 5.1.25 that if $X = X_1 \cup X_2 \cup \cdots$ is a union of path components, then $H_i(X) = \bigoplus_k H_i(X_k)$. Thus it satisfies to prove the theorem in case X is path connected, so we make that assumption.

A singular 0-simplex of X is $f(*) = x$ for some $x \in X$, so we may identify $C_0(X)$ with the free abelian group on the points of X, $C_0(X) = \{\sum_i n_i x_i \mid n_i \in \mathbb{Z}, \; x_i \in X\}$. Since $C_{-1}(X) = 0$, $Z_0(X) = C_0(X)$, i.e., every chain is a cycle. Let $\varepsilon : Z_0(X) \to \mathbb{Z}$ by $\varepsilon(\sum_i n_i x_i) = \sum_i n_i$. We claim that ε is a surjection with kernel $B_0(X)$. Then $H_0(X) = Z_0(X)/B_0(X) \cong \mathbb{Z}$.

Now to prove the claim. First, ε is obviously surjective: Choose $x \in X$. Then for any n, $\varepsilon(nx) = n$. Next, $\mathrm{Ker}(\varepsilon) \supseteq B_0(X)$: $B_0(X)$ is generated by the boundaries of singular 1-simplices. But a singular 1-simplex is a map $f : I \to X$, and the boundary of that is $q - p$ where $q = f(1)$ and $p = f(0)$. Then $\varepsilon(q - p) = 1 - 1 = 0$. Finally, $B_0(X) \subseteq \mathrm{Ker}(\varepsilon)$: Suppose $\varepsilon(\sum_i n_i x_i) = 0$, i.e., $\sum_i n_i = 0$. Rewrite $n_i x_i$ as $x_i + \cdots + x_i$, where there are n_i terms, if $n_i > 0$, or as $-x_i - \cdots - x_i$, where there are $|n_i|$ terms, if $n_i < 0$. Then $\sum_i n_i x_i = x_1' + \cdots + x_k' + (-x_1'') + \cdots + (-x_k'')$ for some k and some points $x_1', \ldots, x_k''$. But now for each i between 1 and k, let c_i be the singular 1-simplex given by $f : I \to X$ with $f(0) = x_i''$ and $f(1) = x_i'$. Then $\partial c_i = x_1' - x_1''$ so $\sum_i n_i x_i = \partial(\sum_{j=1}^k c_j) \in B_0(X)$.   □

Lemma 5.2.2. *Let $f : I \to X$ and $g : I \to X$ with $f(1) = g(0)$. Define $h : I \to X$ by $h(t) = f(2t)$ for $0 \le t \le \frac{1}{2}$, and $h(t) = g(2t - 1)$ for $\frac{1}{2} \le t \le 1$. Then $[f + g - h] = 0 \in H_1(X)$.*

Proof. We exhibit a 2-cell C with $\partial C = f + g - h$. $C : I \to I \to X$ is given by following f and then g along each of the heavy solid lines as indicated:

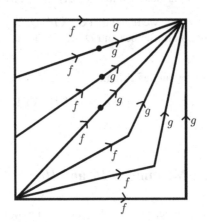

Then $\partial C = f + g - h - k$, where k is the path on the left-hand side of the square. But that is the constant path at $f(0)$, and hence a degenerate 1-chain, so $\partial C = f + g - h$ in $Z_1(X)$.   □

Remark 5.2.3. Obviously this generalizes to the composition of any finite number of paths (proof by induction).   ◊

Theorem 5.2.4. *Let X be a path-connected space. The map $\theta : \pi_1(X, x_0) \to H_1(X)$ given by $\theta(f) = [f]$, where $f : (S^1, 1) \to (X, x_0)$, is an epimorphism with kernel the commutator subgroup of $\pi_1(X, x_0)$. Thus $H_1(X)$ is isomorphic to the abelianization of $\pi_1(X, x_0)$.*

Proof. There are several things to show:

(1) θ is a homomorphism: That follows immediately from Lemma 5.2.2.
(2) θ is surjective: Once and for all, for every point $x \in X$ choose a path α_x from x_0 to x. We make this choice completely arbitrarily, except that we let α_{x_0} be the constant path at x_0. Let β_x be α_x run backwards, β_x from x to x_0.
    Let $z = \sum_{i \in I} a_i T^i$ represent an element of $H_1(X)$, $T^i : I^1 \to X$. Let $p_i = (T^i)^{-1}(0)$ and $q_i = (T^i)^{-1}(1)$. Then $0 = \partial z = \sum_{i \in I} a_i(B_1 T - A_1 T)$ so after gathering terms the coefficient of every 0-simplex $S_j : I^0 \to X$, $j \in J$, is zero. Then $\sum_{i \in I} a_i(\alpha_{p_i} + T^i + \beta_{q_i}) = \sum a_i T^i = z$. Now for each i, $\alpha_{p_i} + T^i + \beta_{q_i}$ is homologous, again by Lemma 5.2.2, to the image of an element of $\pi_1(X, x_0)$,

that element being obtained by beginning at x_0, following α_{p_i} from x_0 to p_i, then following T^i from p_i to q_i, then following β_{q_i} from q_i back to x_0. (Observe that this composite path is a loop at x_0.)

(3) $\mathrm{Ker}(\theta) \supseteq G$, the commutator subgroup of $\pi_1(X,x_0)$. Algebraically, that is immediate, as $\mathrm{Im}(\theta) \subseteq H_1(X)$, which is an abelian group. But geometrically that is easy to see as well. It amounts to showing that if f_1 and f_2 are conjugate elements of $\pi_1(X,x_0)$, then $\theta(f_2) = \theta(f_1)$. But f_1 and f_2 conjugate simply means $f_2 = gf_1g^{-1}$ for some $g \in \pi_1(X,x_0)$, and then $\theta(f_2) = \theta(f_1)$ again by Lemma 5.2.2.

(4) $\mathrm{Ker}(\theta) \subseteq G$. For a 1-cell T, let $\delta(T) = \alpha_p T \beta_q : I \to X$ where $p = T(0)$ and $q = T(1)$. Observe that if T is degenerate, $\delta(T)$ represents 1 in $\pi_1(X,x_0)$. For a 2-cell U, let $\triangle(U) = \delta(A_2U)\delta(B_1U)\delta((B_2U)^{-1})\delta((A_1U)^{-1})$ where the inverse denotes that the path is traversed in the opposite direction.

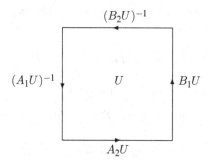

Note that $\triangle(U)$ is homotopic to $\alpha_p \beta_p$, $p = U(0,0)$, so $\triangle(U)$ represents 1 in $\pi_1(X,x_0)$.

Now suppose $\theta(f) = 0$ in $H_1(X)$. Then

$$\theta(f) = \partial \left(\sum_{n \in K} a_k U^k \right) \quad \text{in } C_*(X)$$

$$= \partial \left(\sum_{k \in K} a_k U^k \right) + \sum_{q \in Q} b_q D_q \quad \text{in } Q_*(X)$$

where $\{D_q\}$ are degenerate 1-cells,

$$= \sum a_k (B_1 T^k - A_1 T^k + A_2 T^k - B_2 T^k) + \sum b_q D^q.$$

In this sum, $\theta(f)$ appears with coefficient 1 and every other non-degenerate cell appears with coefficient 0.

Now, if [ ] denotes the homotopy class in $\pi_1(X, x_0)$,

$$\prod_k [\triangle(U^k)]^{a_k} = 1, \quad \prod_q [\delta(D_q)]^{b_q} = 1,$$

so in $\pi_1(X, x_0)/G$,

$$0 = \sum a_k \triangle(U^k) + \sum b_q \delta(D^q)$$
$$= \sum a_k (\delta(B_1 U^k) - \delta(A_1 U^k) + \delta(A_2 U^k) - \delta(B_2 U^k)) + \sum_q b_q \delta(D^q).$$

Applying δ to the expression for $\theta(f)$, and comparing it with the second expression, we see that $\delta(\theta(f))$ represents 0 in $\pi_1(X, x_0)/G$. But $\delta(\theta(f)) = \alpha_{x_0} f \beta_{x_0}$ is homotopic rel $\{0, 1\}$ to f as α_{x_0} is the constant path. Hence $f = 0$ in $\pi_1(X, x_0)/G$.   □

Corollary 5.2.5. *Let X be a path-connected space. The map θ induces a bijection (of sets)*

$$\{\text{free homotopy classes of maps: } S^1 \longrightarrow X\} \longrightarrow H_1(X).$$

Proof. Immediate from Theorems 5.2.4 and 2.5.1.   □

Lemma 5.2.6. *Let $f : (X, x_0) \to (Y, y_0)$. Then the following diagram commutes:*

$$
\begin{array}{ccc}
\pi_1(X, x_0) & \xrightarrow{\;\theta\;} & H_1(X) \\
{\scriptstyle f_*} \downarrow & & \downarrow {\scriptstyle f_*} \\
\pi_1(Y, y_0) & \xrightarrow{\;\theta\;} & H_1(Y).
\end{array}
$$

Using the results of this section, and our work on covering spaces, we now provide an alternate proof of Theorem 4.2.31.

Theorem 5.2.7. *Let d be any integer. Then for any integer $n \geq 1$, there exists a map $f : S^n \to S^n$ of degree d.*

Proof. Again the key step is the $n = 1$ case, and we provide an alternate proof of that (with the remainder of the proof being the same as in the previous proof).

Again we claim that $f : S^1 \to S^1$ by $f(z) = z^d$ has degree d.

To prove that, we consider the diagram

$$
\begin{array}{ccc}
\pi_1(S^1,1) & \xrightarrow{\;\;\theta\;\;} & H_1(S^1) \\
\Big\downarrow{\scriptstyle f_*} & & \Big\downarrow{\scriptstyle f_*} \\
\pi_1(S^1,1) & \xrightarrow{\;\;\theta\;\;} & H^1(S^1).
\end{array}
$$

The horizontal maps are all isomorphisms, as we have shown that $\pi_1(S^1,1) \cong \mathbb{Z}$ in Example 2.2.7. Thus we need only show that $f_* : \pi_1(S^1,1) \to \pi_1(S^1,1)$ is multiplication by d, and to show that it suffices to show that if γ is a generator of $\pi_1(S^1,1)$, then $f_*(\gamma) = d\gamma$. But we also showed that in Example 2.2.7. (The identity map $i : S^1 \to S^1$ represents γ, and then the composition $f \circ i = f$ represents $f_*(\gamma)$.) $\square$

Example 5.2.8. Here is a pair of examples to show that the condition closure $(U) \subseteq$ interior (A) cannot in general be relaxed to $U \subseteq$ interior (A) for the inclusion $(X - U, A - U) \to (X,A)$ to be excisive. In each case we will have a closed set A and we will let $U =$ interior (A), so that we are considering the inclusion $(X\text{-interior}(A), \partial A) \to (X,A)$.

(a) Let $X = \mathbb{R}^2$ and let A be the subset of $\mathbb{R}^2$ that is on or below the graph of the function

$$
f(x) = \begin{cases} \sin\left(\dfrac{1}{x}\right) & x > 0 \\ 1 & x \leq 0. \end{cases}
$$

    Note that ∂A consists of the union of the graph of this function and the interval $\{[0,y] \mid -1 \leq y \leq 1\}$. Observe that both X and X-interior (A) are contractible, that A is path-connected, and that ∂A has two path components.

    Then the exact sequence of the pair (X,A) shows that $H_1(X,A) = 0$, while the exact sequence of the pair $(X\text{-interior}(A), \partial A)$ shows that $H_1(X\text{-interior}(A), \partial A) \cong \mathbb{Z}$.

(b) Let $X = K$, the space of Example 2.6.2, and let $A = K_2$, as in that example. Then interior $(A) = A - (0,0,0)$, and $\partial A = \{(0,0,0)\}$. Then $(X\text{-interior}(A), \partial A) = (K_1, \{(0,0,0)\})$ and so $H_1(X\text{-interior}(A), \partial A) = 0$ as K_1 is contractible and $\{(0,0,0)\}$ is a single point. On the other hand, $H_1(X,A) = H_1(K,K_2) \cong H_1(K)$ by the exact sequence of the pair, as K_2 is contractible. By Corollary 5.2.5 $H_1(K) = \{$free homotopy classes of maps: $S^1 \to K\}$. But this set is nonzero (and in fact $H_1(K)$ is infinitely generated) with a nonzero element being the loop described in Example 2.6.2. $\Diamond$

5.3   Homology with Coefficients

Heretofore we have been considering singular homology with integral coefficients. We now consider arbitrary coefficients. The development of the theory with arbitrary coefficients involves a number of algebraic constructions but no new geometry.

Throughout this section we let G be an abelian group.

Definition 5.3.1. Let (X,A) be a pair. Then

$$Q_n(X;G) = Q_n(X) \otimes G \quad \text{(and similarly for } A)$$
$$D_n(X;G) = D_n(X) \otimes G \quad \text{(and similarly for } A)$$
$$C_n(X;G) = Q_n(X;G)/D_n(X;G) \quad \text{(and similarly for } A)$$
$$C_n(X,A;G) = C_n(X;G)/C_n(A;G). \qquad \qquad \diamond$$

Here we regard the abelian group G as a $\mathbb{Z}$-module, and $\otimes G$ is an abbreviation for $\otimes_{\mathbb{Z}} G$.

In concrete terms, we have for example, that $q \in Q_n(X;G)$ is

$$q = \sum_i g_i \Phi_i$$

where $g_i \in G$ and $\Phi_i : I^n \to X$ is a singular n-cube.

Lemma 5.3.2. $C_n(X;G)$ *is isomorphic to* $C_n(X) \otimes G$ *(and similarly for A). Also,* $C_n(X,A;G)$ *is isomorphic to* $C_n(X,A) \otimes G$.

Proof. Clear from Definition 5.3.1 and the fact that for any two abelian groups A and B, $(A \oplus B) \otimes G \approx (A \otimes G) \oplus (B \otimes G)$, and hence, in this situation, $(A \otimes G) \approx ((A \oplus B) \otimes G)/(B \otimes G)$. $\qquad \square$

Lemma 5.3.3. *With the above identifications,* $C_n(X;G)$ *is a chain complex with boundary map* $\partial \otimes 1 : C_n(X;G) \to C_{n-1}(X;G)$, *and similarly for* $C_n(A;G)$ *and* $C_n(X,A;G)$.

Proof. The only thing to check is that $(\partial \otimes 1)^2 = 0$. But $(\partial \otimes 1)^2 = \partial^2 \otimes 1 = 0$. $\quad \square$

Lemma 5.3.4. *There is a split short exact sequence*

$$0 \longrightarrow C_n(A;G) \longrightarrow C_n(X;G) \longrightarrow C_n(X,A;G) \longrightarrow 0,$$

and hence $C_n(X;G)$ *is isomorphic to* $C_n(A;G) \oplus C_n(X,A;G)$.

Proof. We have the short exact sequence

$$0 \longrightarrow C_n(A) \longrightarrow C_n(X) \longrightarrow C_n(X,A) \longrightarrow 0.$$

Tensoring such a sequence with G does not in general produce an exact sequence. But if this sequence is split short exact, tensoring with G produces a split short exact sequence. This is the case, by Lemma 5.1.26.   □

Lemma 5.3.5. *With the identification in Lemma 5.3.2, $f : X \to Y$ induces $f_* \otimes 1 : C_*(X;G) \to C_*(Y;G)$, and similarly for $f : (X,A) \to (Y,B)$.*

In concrete terms, if $\{\Phi_i : I^n \to X\}$ are singular n-cubes, $(f_* \otimes 1)(\sum_i g_i \Phi_i) = \sum g_i(f\Phi_i)$.

Definition 5.3.6. The *singular homology of X with coefficients in G* is the homology of the chain complex $\{C_*(X;G)\}$. We denote the n-th homology group of this chain complex by $H_n(X;G)$, and similarly for a pair (X,A).   ◇

Theorem 5.3.7. *Singular homology with coefficients in G is an ordinary homology theory with coefficient group G.*

Proof. First we check Axiom 7, the dimension axiom. If X consists of a single point, then $C_*(X;G)$ is isomorphic to

$$\cdots \longrightarrow 0 \longrightarrow 0 \longrightarrow G \longrightarrow 0 \longrightarrow 0 \longrightarrow \cdots$$

with homology as claimed.

The proof that this theory satisfies Axioms 1–3, 5, and 6 is identical to the previous proof.

As for Axiom 4, once we have Lemma 5.3.4, that also follows as before.   □

We have previously denoted singular cohomology with integer coefficients by $H_*(X)$, and we will continue to use that notation, but we will sometimes also denote it by $H_*(X;\mathbb{Z})$ when we wish to emphasize the coefficient group.

Our next goal is to see how to compute $H_*(X;G)$ from $H_*(X;\mathbb{Z})$.

Lemma 5.3.8. *The map $\tau : C_n(X) \to C_n(X;G)$ given by $\tau(\Phi) = \Phi \otimes 1$ where Φ is a singular n-cube induces a map*

$$\tau : H_n(X) \otimes G \longrightarrow H_n(X;G).$$

Proof. Lemma 5.3.3 implies that $\tau : Z_n(X) \to Z_n(X;G)$ and $\tau : B_n(X) \to B_n(X;G)$, where, as usual, $Z_n(X) = \text{Ker}(\partial_n)$ and $B_n(X) = \text{Im}(\partial_{n+1})$, and also $Z_n(X;G) = \text{Ker}(\partial_n \otimes 1)$ and $B_n(X;G) = \text{Im}(\partial_{n+1} \otimes 1)$.   □

Theorem 5.3.9 (Universal coefficient theorem). (1) *For any space X and abelian group G, there is a split short exact sequence*

$$0 \longrightarrow H_n(X) \otimes G \overset{\tau}{\longrightarrow} H_n(X;G) \longrightarrow \text{Tor}(H_{n-1}(X),G) \longrightarrow 0.$$

(2) *If $f : X \to Y$ is a map, there is a commutative diagram*

$$
\begin{array}{ccccccccc}
0 & \longrightarrow & H_n(X) \otimes G & \longrightarrow & H_n(X;G) & \longrightarrow & \mathrm{Tor}\,(H_{n-1}(X),G) & \longrightarrow & 0 \\
& & \downarrow & & \downarrow & & \downarrow & & \\
0 & \longrightarrow & H_n(Y) \otimes G & \longrightarrow & H_n(Y;G) & \longrightarrow & \mathrm{Tor}\,(H_{n-1}(Y),G) & \longrightarrow & 0.
\end{array}
$$

Proof. This is a purely algebraic fact about the homology of chain complexes, and we omit the proof.   □

Remark 5.3.10. This (omitted) proof *crucially* uses the fact that the singular chain complex of a space consists of *free* abelian groups.   ◊

Remark 5.3.11. We have not completely defined Tor here. But we can draw several consequences. First, we always have that the map $H_n(X) \otimes G \to H_n(X;G)$ is an injection, and furthermore that $H_n(X;G) \approx H_n(X) \otimes G \oplus \mathrm{Tor}\,(H_{n-1}(X),G)$. In particular, if $H_{n-1}(X) = 0$, or, by Lemma A.3.8, if $H_{n-1}(X)$ is free abelian, then $H_n(X;G) \approx H_n(X) \otimes G$.

Thus, if X is a space with $H_n(X)$ torsion-free for all n, then $H_n(X;G) \approx H_n(X) \otimes G$ for every n. Examples of these are spheres (by Lemma 4.1.3) or complex projective spaces (by Theorem 4.3.3).   ◊

Example 5.3.12. By Lemma A.3.8, $\mathrm{Tor}\,(\mathbb{Z}_2, \mathbb{Z}_m) \approx \mathbb{Z}_2$ for m even. Thus, from Theorem 4.3.4, for real projective spaces we have, for m even,

$$
H_i(\mathbb{R}P^n; \mathbb{Z}_m) = \begin{cases}
0 & i > n \\
\mathbb{Z}_m & i = n \text{ odd} \\
\mathbb{Z}_2 & i = n \text{ even} \\
\mathbb{Z}_2 & 1 \le i \le n-1 \\
\mathbb{Z}_m & i = 0.
\end{cases}
$$

On the other hand, by Lemma A.3.8, $\mathrm{Tor}\,(\mathbb{Z}_2, \mathbb{Z}_m) = 0$ for m odd. Thus for real projective spaces we have, for m odd

$$
H_i(\mathbb{R}P^n; \mathbb{Z}_m) = \begin{cases}
0 & i > n \\
\mathbb{Z}_m & i = n \text{ odd} \\
0 & i = n \text{ even} \\
0 & 1 \le i \le n-1 \\
\mathbb{Z}_m & i = 0.
\end{cases}
$$
                                                                                                                                     ◊

Corollary 5.3.13. *Let $f : X \to Y$ and suppose that $f_* : H_n(X) \to H_n(Y)$ is an isomorphism for all n. Then $f_* : H_n(X;G) \to H_n(Y;G)$ is an isomorphism for all n.*

Proof. This follows directly from the universal coefficient theorem and the short five lemma. □

Remark 5.3.14. Once we have that $H_n(X;G)$ is a homology theory, we have all the consequences that follow from the axioms, e.g., the Mayer-Vietoris sequence. We also have the cellular homology with coefficients in G of a CW-complex, and again it is isomorphic to singular homology with coefficients in G. ◇

Remark 5.3.15. If we tensor with a commutative ring R, instead of a group G, then $C_n(X;R)$ has the structure of an R-module, and hence so does $H_n(X;R)$. An important special case is that of a field $\mathbb{F}$, whence $H_n(X;\mathbb{F})$ is an $\mathbb{F}$-vector space. ◇

We can also ask what happens if we change coefficient groups, and the answer is what we would expect.

Theorem 5.3.16. *Let $\varphi : G_1 \to G_2$ be a homomorphism. Then there is a commutative diagram of split short exact sequences, with all vertical maps induced by φ,*

$$
\begin{array}{ccccccccc}
0 & \longrightarrow & H_n(X) \otimes G_1 & \longrightarrow & H_n(X;G_1) & \longrightarrow & \mathrm{Tor}\,(H_{n-1}(X),G_2) & \longrightarrow & 0 \\
& & \downarrow & & \downarrow & & \downarrow & & \\
0 & \longrightarrow & H_n(X) \otimes G_2 & \longrightarrow & H_n(X;G_2) & \longrightarrow & \mathrm{Tor}\,(H_{n-1}(X),G_2) & \longrightarrow & 0.
\end{array}
$$

5.4 The Künneth Formula

Our goal in this section is to derive the Künneth formula, which expresses the homology of a product $X \times Y$ in terms of the homology of each of the factors X and Y.

Definition 5.4.1. Let $\Phi : I^j \to X$ be a singular j-cube and $\Psi : I^k \to Y$ be a singular k-cube. Then their *cross product* $\Phi \times \Psi : I^j \times I^k \to X \times Y$ is the singular $(j+k)$-cube given by

$$(\Phi \times \Psi)(x_1,\ldots,x_j,y_1,\ldots,y_k) = (\Phi(x_1,\ldots,x_j),\Psi(y_1,\ldots,y_k)).$$

◇

Lemma 5.4.2. *The cross product induces a map*

$$C_j(X) \otimes C_k(Y) \longrightarrow C_{j+k}(X \times Y).$$

Proof. If either Φ or Ψ is degenerate, so is $\Phi \times \Psi$. □

Lemma 5.4.3. *In this situation,*

$$\partial(\Phi \times \Psi) = (\partial\Phi) \times \Psi + (-1)^j \Phi \times (\partial\Psi).$$

Proof. Direct calculation, with careful attention to signs. □

By definition, the tensor product of the chain complex $\mathscr{A} = \{A_i\}$ and $\mathscr{B} = \{B_j\}$ is $\mathscr{C} = \{C_k\}$ where $C_k = \bigoplus_{i+j=k} A_i \otimes B_j$. Thus, given this definition, we have a map from $C_*(X) \otimes C_*(Y) \to C_*(X \times Y)$.

We now cite the Eilenberg-Zilber theorem:

Theorem 5.4.4. *The map*

$$C_*(X) \otimes C_*(Y) \longrightarrow C_*(X \times Y)$$

induces an isomorphism on homology.

Once we have this theorem, we may use algebraic methods to compute the homology of $X \times Y$ from the homology of X and the homology of Y. But the first thing to note is that the homology of $C_*(X) \otimes C_*(Y)$ is not in general isomorphic to $H_*(X) \otimes H_*(Y)$, though they are closed related, as we shall now see.

Lemma 5.4.5. *The cross product induces a map*

$$H_j(X) \otimes H_k(Y) \longrightarrow H_{j+k}(X \times Y).$$

Proof. First we show that we obtain a map

$$Z_j(X) \otimes Z_k(Y) \longrightarrow Z_{j+k}(X \times Y).$$

Let $c \in Z_j(X)$ and $d \in Z_k(Y)$ be singular cycles, so that $\partial c = 0$ and $\partial d = 0$. Then

$$\partial(c \times d) = (\partial c) \times d + (-1)^j c \times (\partial d) = 0 \times d \pm c \times 0 = 0$$

so $c \times d \in Z_{j+k}(X, Y)$.

Now suppose $c \in B_j(X)$, $c = \partial e$ for some $e \in C_{j+1}(X)$. Then $c \times d = \partial(e \times d)$, so $c \times d \in B_{j+k}(X \times Y)$. Similarly if $d \in B_k(Y)$, $c \times d \in B_{j+k}(X \times Y)$. Thus we obtain a map

$$(Z_j(X)/B_j(X)) \otimes (Z_k(Y)/B_k(Y)) \longrightarrow Z_{j+k}(X \times Y)/B_{j+k}(X \times Y),$$

i.e., a map

$$H_j(X) \otimes H_k(Y) \longrightarrow H_{j+k}(X \times Y).$$ □

We regard $H_*(X)$ and $H_*(Y)$, and hence $H_*(X) \otimes H_*(Y)$, as chain complexes with identically zero boundary operator.

We then have

Theorem 5.4.6 (Künneth formula). (1) *For any spaces X and Y, there is a split short exact sequence*

$$0 \longrightarrow (H_*(X) \otimes H_*(Y))_n \longrightarrow H_n(X \times Y) \longrightarrow (\mathrm{Tor}\,(H_*(X), H_*(Y)))_{n-1} \longrightarrow 0.$$

(2) *For any spaces X, Y, Z, W, and maps* $f : X \to Z$, $g : Y \to W$, *there is a commutative diagram*

$$
\begin{array}{ccccccccc}
0 & \longrightarrow & (H_*(X) \otimes H_*(Y))_n & \longrightarrow & H_n(X \times Y) & \longrightarrow & (\mathrm{Tor}\,(H_*(X), H_*(Y)))_{n-1} & \longrightarrow & 0 \\
& & \downarrow & & \downarrow & & \downarrow & & \\
0 & \longrightarrow & (H_*(Z) \otimes H_*(Y))_n & \longrightarrow & H_n(Z \times W) & \longrightarrow & (\mathrm{Tor}\,(H_*(Z), H_*(W)))_{n-1} & \longrightarrow & 0.
\end{array}
$$

In the statement of this theorem,

$$(H_*(X) \otimes H_*(Y))_n = \bigoplus_{j=0}^{n} H_j(Y) \otimes H_{n-j}(Y),$$

$$(\mathrm{Tor}\,(H_*(X), H_*(Y)))_{n-1} = \bigoplus_{j=0}^{n-1} \mathrm{Tor}\,(H_j(Y), H_{n-1-j}(Y)).$$

Proof. This is a purely algebraic result, whose proof we omit, but we again remark that it crucially uses the fact that $C_*(X)$ and $C_*(Y)$ are chain complexes of *free* abelian groups. □

Corollary 5.4.7. *Let* $\mathbb{F}$ *be a field. Then for any spaces X and Y, the map*

$$(H_*(X; \mathbb{F}) \otimes H_*(Y; \mathbb{F}))_n \longrightarrow H_n(X \times Y; \mathbb{F})$$

is an isomorphism.

Recall that for a path connected space Y, we have a canonical choice of $1 \in H_0(Y)$. For clarity we shall denote this element by 1_Y. Let us identify $H_n(X) \otimes H_0(Y)$ with a subgroup of $H_n(X \times Y)$ via the inclusion in the Künneth formula.

Lemma 5.4.8. *Let Y be a path connected space and let* $\pi : X \times Y \to X$ *be projection on the first factor. For any element* α *of* $H_n(X)$,

$$\pi_*(\alpha \otimes 1_Y) = \alpha.$$

Proof. Clear from the construction in Lemma 5.4.5. □

We have stated the Künneth formula for spaces. A similar formula holds for pairs, where by definition

$$(X,A) \times (Y,B) = (X \times Y, X \times B \cup Y \times A).$$

The proof is very much along the same lines, and again we must apply a form of the Eilenberg-Zilber theorem for pairs. But in this situation this theorem requires a slight additional hypothesis.

Theorem 5.4.9. *Suppose that $\{X \times B, A \times Y\}$ is an excisive couple. Then the map*

$$C_*(X,A) \otimes C_*(Y,B) \longrightarrow C_*((X,A) \times (Y,B))$$

induces an isomorphism on homology.

This yields the Künneth formula under the same additional hypothesis.

Theorem 5.4.10 (Künneth formula). (1) *For any pairs (X,A) and (Y,B) with $\{X \times B, A \times Y\}$ an excisive couple in $X \times Y$, there is a split short exact sequence*

$$0 \longrightarrow (H_*(X,A) \otimes H_*(Y,B))_n \longrightarrow H_n((X,A) \times (Y,B))$$
$$\longrightarrow (\operatorname{Tor}(H_*(X,A), H_*(Y,B)))_{n-1} \longrightarrow 0.$$

(2) *For any pairs (X,A) and (Y,B) with $\{X \times B, A \times Y\}$ an excisive couple in $X \times Y$, and any pairs (Z,C) and (W,D) with $\{Z \times D, C \times W\}$ an excisive couple in $Z \times W$, and any maps of pairs $f : (X,A) \to (Z,C)$ and $g : (Y,B) \to (W,D)$, there is a commutative diagram*

$$
\begin{array}{ccccccc}
0 \to & (H_*(X,A) \otimes H_*(Y,B))_n & \to & H_n((X,A) \times (Y,B)) & \to & (\operatorname{Tor}(H_*(X,A), H_*(Y,B)))_{n-1} & \to 0 \\
& \downarrow & & \downarrow & & \downarrow & \\
0 \to & (H_*(Z,C) \otimes H_*(W,D))_n & \to & H_n((Z,C) \times (W,D)) & \to & (\operatorname{Tor}(H_*(Z,C), H_*(W,D)))_{n-1} & \to 0.
\end{array}
$$

Corollary 5.4.11. *Let $\mathbb{F}$ be a field. Then for any pairs (X,A) and (Y,B) such that $\{X \times B, A \times Y\}$ is an excisive couple, the map*

$$(H_*(X,A;\mathbb{F}) \otimes H_*(Y,B;\mathbb{F}))_n \longrightarrow H_n((X,A) \times (Y,B);\mathbb{F})$$

is an isomorphism.

5.5   Cohomology

In this section we develop singular cohomology. Given our previous work, this is almost entirely algebraic. Then we extend it and relate it to other things we have done.

Definition 5.5.1. The *singular cohomology (with integer coefficients)* $H^*(X,A)$ is the homology of the dual cochain complex of the integral singular chain complex $C_*(X,A)$.   ◊

We now elaborate on this definition.

Given the singular chain complex $C_*(X)$, we form the dual cochain complex $C^*(X)$ given by

$$C^n(X) = \mathrm{Hom}\,(C_n(X),\mathbb{Z})$$

with coboundary operator $\delta^n : C^n(X) \to C^{n+1}(X)$ given by

$$\delta^n(\gamma)(c) = \gamma(\partial_{n+1}c)$$

where $\gamma \in C^n(X)$ and $c \in C_{n+1}(X)$. The relation $\partial_{n-1}\partial_n = 0$ immediately yields $\delta^n\delta^{n-1} = 0$ as well.

Definition 5.5.2. The group of *singular n-cocycles* $Z^n(X)$ and the group of *singular n-coboundaries* $B^n(X)$ are defined by:

$$Z^n(X) = \mathrm{Ker}\,(\delta^n : C^n(X) \longrightarrow C^{n+1}(X))$$
$$B^n(X) = \mathrm{Im}\,(\delta^{n-1} : C^{n-1}(X) \longrightarrow C^n(X)).$$

The *n*-th *singular cohomology group* of X is defined by

$$H^n(X) = Z^n(X)/B^n(X).$$ ◊

Let $f : X \to Y$ be a map of spaces. Then f induces a map $f^* : C^*(Y) \to C^*(X)$ by

$$f^*(\gamma)(c) = \gamma(f(c))$$

where $\gamma \in C^n(Y)$ and $c \in C_n(X)$.

Lemma 5.5.3. *The map* $f^* : C^*(Y) \to C^*(X)$ *induces a map* $f^* : H^*(Y) \to H^*(X)$.

Proof. It is routine to check that $f^*(Z^n(Y)) \subseteq Z^n(X)$ and $f^*(B^n(Y)) \subseteq B^n(X)$. □

Theorem 5.5.4. *Singular cohomology is an ordinary cohomology theory with* $\mathbb{Z}$ *coefficients.*

Proof. This proof entirely mimics the proof that singular homology is an ordinary homology theory with $\mathbb{Z}$ coefficients. □

There is just one subtlety, Axiom 4, the exactness axiom. Exactness for homology followed from the short exactness of the sequence of singular chain complexes

$$0 \longrightarrow C_*(A) \longrightarrow C_*(X) \longrightarrow C_*(X,A) \longrightarrow 0,$$

i.e., the exactness of

$$0 \longrightarrow C_n(A) \longrightarrow C_n(X) \longrightarrow C_n(X,A) \longrightarrow 0.$$

In general, however, if $0 \to A \to B \to C \to 0$ is a sequence of $\mathbb{Z}$-modules, $0 \to C^* \to B^* \to A^* \to 0$ is not exact (where A^*, B^*, and C^* are the duals of A, B, and C respectively).

However, if $0 \to A \to B \to C \to 0$ is split exact then $0 \to C^* \to B^* \to A^* \to 0$ is exact, and in fact split exact.

But that is the situation we are in here. The sequence

$$0 \longrightarrow C_*(A) \longrightarrow C_*(X) \longrightarrow C_*(X,A) \longrightarrow 0$$

is split exact by Lemma 5.1.26.

There is also one useful trick. We have not given the proof of axiom 5, the excision axiom, for homology, due to its length. One can prove excision for cohomology along the same lines as the proof for homology. But instead, using Corollary 5.5.13 below, we can calculate that excision for homology immediately implies excision for cohomology.

Remark 5.5.5. In the construction of cohomology, there is one algebraic subtlety we need to note. Let M be a free abelian group of finite rank. Then $M^* = \mathrm{Hom}(M, \mathbb{Z})$ is a free abelian group of the same rank. But when we leave the finite rank case we run into problems. For example, if M is a free abelian group of countably infinite rank, then M^* is an uncountable torsion-free abelian group that is *not* free. (This is a theorem of Baer.)

The infinite rank case is precisely the situation we are in here, as if X is any space other than a finite set of points, then the singular chain complex $C_*(X)$ is infinitely generated. To deal with this situation we will have to impose various finiteness assumptions in several places below. $\Diamond$

Having obtained singular cohomology with integer coefficients, we may then obtain singular cohomology with coefficient group G in exactly the same way we obtained singular homology with coefficient group G.

Definition 5.5.6. Let G be an abelian group. The *singular cohomology of X with coefficients G*, $H^*(X;G)$, is the homology of $C^*(X) \otimes G$, where $C^*(X)$ is the dual chain complex to the singular chain complex of X. $\Diamond$

Theorem 5.5.7. *Singular cohomology with coefficients in G is an ordinary cohomology theory with coefficient group $G = H^0(X;G)$.*

Proof. Again this mirrors the proof for singular homology in Sect. 5.1. Again Axiom 4 uses the fact that, for every n, the sequence $0 \to C^n(X,A) \to C^n(X) \to C^n(A) \to 0$ is split exact. $\square$

Again we have a universal coefficient theorem for cohomology. Because of the problem with infinite ranks, we need additional hypotheses.

Theorem 5.5.8 (Universal coefficient theorem). (1) *Let X be a space and let G be an abelian group. Suppose that X is of finite type or that G is of finitely generated. Then there is a split short exact sequence*

$$0 \longrightarrow H^n(X) \otimes G \longrightarrow H^n(X;G) \longrightarrow \mathrm{Tor}(H^{n+1}(X),G) \longrightarrow 0.$$

(2) *In this situation, if $f : X \to Y$ is a map, there is a commutative diagram*

$$
\begin{array}{ccccccccc}
0 & \longrightarrow & H^n(Y) \otimes G & \longrightarrow & H^n(Y;G) & \longrightarrow & \mathrm{Tor}(H^{n+1}(Y),G) & \longrightarrow & 0 \\
 & & \downarrow & & \downarrow & & \downarrow & & \\
0 & \longrightarrow & H^n(X) \otimes G & \longrightarrow & H^n(X;G) & \longrightarrow & \mathrm{Tor}(H^{n+1}(X),G) & \longrightarrow & 0.
\end{array}
$$

Proof. Again we omit the purely algebraic argument, but we note that, while the cochain groups $C^*(X)$ are not in general free, they are torsion-free, and that fact, together with our additional hypotheses, suffices to be able to apply that argument. □

We have the following corollary (compare Corollary 5.3.13).

Corollary 5.5.9. *In the situation of Theorem 5.5.8, let $f : X \to Y$ and suppose that $f^* : H^n(Y) \to H^n(X)$ is an isomorphism for all n. Then $f^* : H^n(Y;G) \to H^n(X;G)$ is an isomorphism for all n.*

We also have the analog of Theorem 5.3.16.

Theorem 5.5.10. *In the situation of Theorem 5.5.8, let $\varphi : G_1 \to G_2$ be a homomorphism. Then there is a commutative diagram of split short exact sequences, with all vertical maps induced by φ,*

$$
\begin{array}{ccccccccc}
0 & \longrightarrow & H^n(X) \otimes G_1 & \longrightarrow & H^n(X;G_1) & \longrightarrow & \mathrm{Tor}(H^{n+1}(X),G_1) & \longrightarrow & 0 \\
 & & \downarrow & & \downarrow & & \downarrow & & \\
0 & \longrightarrow & H^n(X) \otimes G_2 & \longrightarrow & H^n(X;G_2) & \longrightarrow & \mathrm{Tor}(H^{n+1}(X),G_2) & \longrightarrow & 0.
\end{array}
$$

It is natural to expect that there will be a close relationship between homology and cohomology, and that is indeed the case. We develop that now.

Given the definition of the dual chain complex, we have the *evaluation map*

$$e : C^n(X) \otimes C_n(X) \longrightarrow \mathbb{Z}$$

given by evaluating a cochain on a chain, i.e.

$$e(\gamma, c) = \gamma(c) \quad \text{for } \gamma \in C^n(X),\ c \in C_n(X).$$

Lemma 5.5.11. *The evaluation map e induces a map*

$$e : H^n(X) \otimes H_n(X) \longrightarrow \mathbb{Z}$$

by $e([\gamma], [c]) = \gamma(c)$, where γ (resp. c) is a representative of the cohomology class $[\gamma]$ (resp. the homology class $[c]$).

Proof. We can restrict e to evaluate cocycles on cycles,

$$e : Z^n(X) \otimes Z_n(X) \longrightarrow \mathbb{Z}$$

by $e(\gamma, c) = \gamma(c)$. But then if c is a boundary, $c = \partial d$, $e(\gamma, c) = e(\gamma, \partial d) = e(\delta \gamma, d) = e(0, d) = 0$, and similarly if γ is a coboundary.   □

Given this lemma, we have a map

$$e : H^n(X) \longrightarrow \mathrm{Hom}\,(H_n(X), \mathbb{Z})$$

given by

$$e([\gamma])([c]) = e([\gamma], [c]) = \gamma(c),$$

i.e., $f = e([\gamma])$ is the homomorphism $f : H_n(X) \to \mathbb{Z}$ given by $f([c]) = \gamma(c)$.

This construction can be performed with arbitrary coefficients, to obtain maps

$$e : H^n(X; G) \otimes H_n(X) \longrightarrow G$$

and

$$e : H^n(X, G) \longrightarrow \mathrm{Hom}\,(H_n(X), G).$$

Theorem 5.5.12 (Universal coefficient theorem). (1) *For any space X and abelian group G, there is a split short exact sequence*

$$0 \longrightarrow \mathrm{Ext}\,(H_{n-1}(X), G) \longrightarrow H^n(X; G) \overset{e}{\longrightarrow} \mathrm{Hom}\,(H_n(X), G) \longrightarrow 0.$$

(2) *If $f : X \to Y$ is a map, there is a commutative diagram*

$$0 \longrightarrow \operatorname{Ext}(H_{n-1}(Y),G) \longrightarrow H^n(Y;G) \longrightarrow \operatorname{Hom}(H_n(Y),G) \longrightarrow 0$$

$$0 \longrightarrow \operatorname{Ext}(H_{n-1}(X),G) \longrightarrow H^n(X;G) \longrightarrow \operatorname{Hom}(H_n(X),G) \longrightarrow 0.$$

Proof. Again this is a purely algebraic argument which we omit. □

Note that this theorem has important consequences even (indeed, especially) in the case $G = \mathbb{Z}$.

Corollary 5.5.13. *Let $f : X \to Y$ induce an isomorphism $f_* : H_n(X) \to H_n(Y)$ for all n. Then $f^* : H^n(Y) \to H^n(X)$ is an isomorphism for all n.*

Corollary 5.5.14. *If the inclusion $(X - U, A - U) \to (X,A)$ is excisive for singular homology, it is excisive for singular cohomology.*

Corollary 5.5.15. *Let X be a space of finite type and suppose that $H_n(X) \approx F_n \oplus T_n$, where F_n is a free abelian group and T_n is a torsion group, for each n. Then*

$$H^n(X) \approx F_n \oplus T_{n-1}$$

for each n.

Proof. This follows from the computation of Ext in Lemma A.3.12. □

Corollary 5.5.16. *Let X be a space of finite type. Then $H^n(X)$ is finitely generated for all n.*

Example 5.5.17. The integral singular cohomology of $\mathbb{R}P^n$ is as follows:

$$H^k(\mathbb{R}P^n) = \begin{cases} 0 & k > n \\ \mathbb{Z} & k = n \text{ odd} \\ 0 & k = n \text{ even} \\ \mathbb{Z}_2 & 1 \le k \le n-1 \text{ even} \\ 0 & 1 \le k \le n-1 \text{ odd} \\ \mathbb{Z} & k = 0, \end{cases}$$

as we see from Corollary 5.5.15 and Theorem 4.3.4. ◊

Remark 5.5.18. We may take $\operatorname{Hom}(\,,R)$ where R is a commutative ring, and then $H^n(X;R)$ becomes an R-module. In particular we may take $R = \mathbb{F}$, a field. We then see that $H_n(X;\mathbb{F})$ and $H^n(X;\mathbb{F})$ are $\mathbb{F}$-vector spaces, and either they are both finite-dimensional vector spaces or they are both infinite-dimensional vector spaces. In case they are both finite-dimensional, then they are dual vector spaces with the pairing

$$e : H^n(X;\mathbb{F}) \otimes H_n(X;\mathbb{F}) \longrightarrow \mathbb{F}$$

being nonsingular.

(In particular, $H^n(X;\mathbb{F})$ and $H_n(X;\mathbb{F})$ are $\mathbb{F}$-vector spaces of the same dimension.) $\Diamond$

We have Theorem 5.5.12, which tells us how to pass from homology to cohomology, and now we have a theorem which tells us how to pass back (under favorable circumstances).

Theorem 5.5.19 (Universal coefficient theorem). (1) *Let X be a space of finite type. For any abelian group G there is a split short exact sequence*

$$0 \longrightarrow \mathrm{Ext}\,(H^{n+1}(X),G) \longrightarrow H_n(X;G) \overset{e}{\longrightarrow} \mathrm{Hom}\,(H^n(X),G) \longrightarrow 0.$$

(2) *If $f : X \to Y$ is a map, where both X and Y are spaces of finite type, there is a commutative diagram*

$$
\begin{array}{ccccccccc}
0 & \longrightarrow & \mathrm{Ext}\,(H^{n+1}(X),G) & \longrightarrow & H_n(X;G) & \longrightarrow & \mathrm{Hom}\,(H^n(X),G) & \longrightarrow & 0 \\
 & & \downarrow & & \downarrow & & \downarrow & & \\
0 & \longrightarrow & \mathrm{Ext}\,(H^{n+1}(Y),G) & \longrightarrow & H_n(Y,G) & \longrightarrow & \mathrm{Hom}\,(H^n(Y),G) & \longrightarrow & 0.
\end{array}
$$

Proof. Again we omit this purely algebraic proof. $\square$

(The reader has undoubtedly observed by now that we have several theorems called the universal coefficient theorem. This reflects the fact that all these theorems have this name in the literature.)

Recall we defined the Euler characteristic $\chi(X)$ of a space X with finitely generated homology in Definition 4.2.19. We observe:

Theorem 5.5.20. *Let X be a space with finitely generated homology. Let $\mathbb{F}$ be an arbitrary field. Then $\chi(X)$ is given by*

$$\chi(X) = \begin{cases} \displaystyle\sum_{n=0}^{\infty} (-1)^n \operatorname{rank} H_n(X;\mathbb{Z}) \\[2ex] \displaystyle\sum_{n=0}^{\infty} (-1)^n \operatorname{rank} H^n(X;\mathbb{Z}) \\[2ex] \displaystyle\sum_{n=0}^{\infty} (-1)^n \dim H_n(X;\mathbb{F}) \\[2ex] \displaystyle\sum_{n=0}^{\infty} (-1)^n \dim H^n(X;\mathbb{F}). \end{cases}$$

Proof. This follows directly from the universal coefficient theorems. (If $\mathbb{F}$ is a field of characteristic zero, then all of these ranks are equal for every integer n. If $\mathbb{F}$ does not have characteristic 0, that may not be the case, depending on the space X, but nevertheless the alternating sums have the same value.)   □

Remark 5.5.21. Recall that we chose a canonical identification of $H_0(*;\mathbb{Z})$ with $\mathbb{Z}$, with $1_* \in H_0(* : \mathbb{Z})$ being identified with $1 \in \mathbb{Z}$.

This gives a canonical identification of $H^0(*;\mathbb{Z})$ with $\mathbb{Z}$ as well: The class $1^* \in H^0(*;\mathbb{Z})$ is the class that evaluates to 1 on $1_* \in H_0(*;\mathbb{Z})$ under the evaluation map of Lemma 5.5.11, i.e., we have the equation $e(1^*, 1_*) = 1$.

Now let X be an arbitrary nonempty space. Then we have the (unique) map $\varepsilon : X \to *$ and hence we have the identical maps ε_* on homology and ε^* on cohomology. Under the above identifications, we may regard $\varepsilon_* : H_0(X;\mathbb{Z}) \to \mathbb{Z}$ and $\varepsilon^* : \mathbb{Z} \to H^0(X;\mathbb{Z})$. In particular we let $1^X \in H^0(X;\mathbb{Z})$ be $\varepsilon^*(1^*)$.

(In case X is path connected, we have the class $1_X \in H_0(X;\mathbb{Z})$ which is the image of 1_* under an arbitrary map $* \to X$, and then we again have $\varepsilon(1^X, 1_X) = 1$.)

We have stated this for $\mathbb{Z}$ coefficients for simplicity but this gives $1^X \in H^0(X)$ for arbitrary coefficients by the universal coefficient theorem.

(We will sometimes abbreviate 1^X to 1, or 1_X to 1, when there is no possibility of confusion.)   ◊

Finally, we have a Künneth formula for cohomology as well as for homology. Again we need finiteness assumptions.

Theorem 5.5.22 (Künneth formula).

(1) *Let X and Y be spaces of finite type. There is a split short exact sequence*

$$0 \longrightarrow (H^*(X) \otimes H^*(Y))^n \longrightarrow H^n(X \times Y) \longrightarrow (\operatorname{Tor}(H^*(X), H^*(Y)))^{n+1} \longrightarrow 0.$$

(2) *For any spaces X, Y, Z, W and maps $f : X \to Z$, $g : Y \to W$, there is a commutative diagram*

$$0 \longrightarrow (H^*(Z) \otimes H^*(W))^n \longrightarrow H^n(Z \times W) \longrightarrow (\text{Tor}\ (H^*(Z), H^*(W)))^{n+1} \longrightarrow 0$$

$$0 \longrightarrow (H^*(X) \otimes H^*(Y))^n \longrightarrow H^n(X \times Y) \longrightarrow (\text{Tor}\ (H^*(X), H^*(Y)))^{n+1} \longrightarrow 0.$$

In the statement of this theorem,

$$(H^*(X) \otimes H^*(Y))^n = \bigoplus_{j=0}^{n} H^j(X) \otimes H^{n-j}(Y),$$

$$(\text{Tor}\ (H^*(X), H_*(Y)))^{n+1} = \bigoplus_{j=0}^{n+1} \text{Tor}\ (H^j(X), H^{n+1-j}(Y)).$$

We identify $H^n(X) \otimes H^0(Y)$ with a subgroup of $H^n(X \times Y)$ by the inclusion in the Künneth formula. We then have the analog in cohomology of Lemma 5.4.8.

Lemma 5.5.23. *Let $\pi : X \times Y \to X$ be projection on the first factor. For any element α of $H^n(X)$,*

$$\pi^*(\alpha) = \alpha \otimes 1^Y.$$

We also have another corollary of the universal coefficient theorem.

Corollary 5.5.24. *Let $\mathbb{F}$ be a field. Then for any spaces X and Y of finite type,*

$$(H^*(X;\mathbb{F}) \otimes H^*(Y;\mathbb{F}))^n \longrightarrow H^n(X \times Y;\mathbb{F})$$

is an isomorphism.

Remark 5.5.25. In exactly the same way that Theorem 5.4.9 generalizes Theorem 5.4.4, and under the exact same additional hypothesis, the Künneth formula for the cohomology of a product of spaces generalizes to the Künneth formula for a product of pairs.   ◊

5.6   Cup and Cap Products

In this section we fix a commutative ring R and an R-algebra M, and assume all of our homology and cohomology groups have coefficients in M. The most important special case is where $R = \mathbb{Z}$ and M is a $\mathbb{Z}$-algebra, i.e., a commutative ring. Of course, this includes the case when M is a field. We suppress the coefficients from our notation.

Our goal in this section is to define the cup product

$$\cup : H^j(X) \otimes H^k(X) \longrightarrow H^{j+k}(X)$$

and the closely related cap product

$$\cap : H^j(X) \otimes H_{j+k}(X) \longrightarrow H_k(X).$$

As we shall see, the cup product gives the cohomology of a space X an algebra structure, or, more precisely, the structure of a commutative graded algebra. It is this extra structure that makes cohomology more powerful than homology.

We begin by revisiting our work from Sect. 5.4, where we defined the cross product in homology. Given one crucial step, a more precise formulation of the Eilenberg-Zilber theorem, we can define the cross product in cohomology in almost exactly the same way. It is then easy to obtain the cup product from the cross product in cohomology, using the contravariance of cohomology. We obtain the cup product by first defining the slant product and then using the covariance of homology.

Again, for simplicity, we will begin by developing the theory for spaces, and then we will state the analogous results for pairs.

We remind the reader of the definition of cross product on chains, Definition 5.4.1, and of Lemmas 5.4.2 and 5.4.5.

We then have the precise version of the Eilenberg-Zilber theorem, Theorem 5.4.4.

Theorem 5.6.1. *There are natural maps of chain complexes*

$$E : C_*(X) \otimes C_*(Y) \longrightarrow C_*(X \times Y)$$

and

$$F : C_*(X \times Y) \longrightarrow C_*(X) \otimes C_*(Y)$$

that are inverse chain equivalences, i.e., that are chain maps that induce inverse isomorphisms on homology

$$E_* : H_*(C_*(X) \otimes C_*(Y)) \longrightarrow H_*(C_*(X \times Y)),$$
$$F_* : H_*(C_*(X \times Y)) \longrightarrow H_*(C_*(X) \otimes C_*(Y)).$$

Here E is the map of Lemma 5.4.2, and we do not define F. With this more precise definition the cross product on chains

$$\times : C_j(X) \otimes C_k(Y) \longrightarrow C_{j+k}(X \times Y)$$

is the map given by

$$(\Phi, \Psi) \longmapsto E(\Phi \times \Psi).$$

Now consider a singular cochain. It is an element of the dual of the space of singular chains, so it is determined by its values on singular chains, and indeed by its values on a basis of the group of singular chains. Now, given

$$f \in C^j(X) \quad \text{and} \quad g \in C^k(X)$$

and

$$\Phi \in C_j(X) \quad \text{and} \quad \Psi \in C_k(X)$$

we have the evaluation map

$$(f,g)(\Phi,\Psi) = f(\Phi)g(\Psi).$$

(Note in taking this product we are using the hypothesis that our coefficients are in an algebra.)

It is immediate that this evaluation map is bilinear, giving a map

$$(C^j(X) \otimes C^k(Y)) \otimes (C_j(X) \otimes C_k(Y)) \longrightarrow M.$$

Recall that $C_j(X) \otimes C_k(Y)$ is one summand in

$$(C_*(X) \otimes C_*(Y))_{j+k} = \bigoplus_{p+q=j+k} C_p(X) \otimes C_q(Y)$$

and we extend this map to

$$\times : (C^j(X) \otimes C^k(Y)) \otimes (C_*(X) \otimes C_*(Y))_{j+k} \longrightarrow M$$

by requiring that it be identically zero on the other summands.

This is almost what we need, and the missing link is provided by the other Eilenberg-Zilber map.

Definition 5.6.2. The cross product on cochains

$$\times : C^j(X) \otimes C^k(Y) \longrightarrow C^{j+k}(X \times Y)$$

is the map given by

$$f \otimes g \longmapsto (f \times g)(F),$$

i.e., if $c \in C_{j+k}(X \times Y)$ is a chain, and $F(c) = \sum m_i \Phi_i \otimes \Psi_i$, then

$$(f \times g)(c) = \sum m_i f(\Phi_i) g(\Psi_i). \qquad \Diamond$$

Lemma 5.6.3. *In this situation,*

$$\delta(f \times g) = (\delta f) \times g + (-1)^j f \times (\delta g).$$

Proof. Entirely analogous to the proof of Lemma 5.4.3.   □

Lemma 5.6.4. *The cross product induces a map*

$$\times : H^j(X) \otimes H^k(Y) \longrightarrow H^{j+k}(X \times Y).$$

Proof. Entirely analogous to the proof of Lemma 5.4.5.   □

We record several properties of both cross products. First we have naturality.

Theorem 5.6.5. *Let* $\alpha : X_1 \to X_2$ *and* $\beta : Y_1 \to Y_2$ *be maps. Then there are commutative diagrams*

$$
\begin{array}{ccc}
H_j(X_1) \otimes H_k(Y_1) & \longrightarrow & H_{j+k}(X_1 \times Y_1) \\
\alpha_* \otimes \beta_* \downarrow & & \downarrow (\alpha \times \beta)_* \\
H_j(X_2) \otimes H_k(Y_2) & \longrightarrow & H_{j+k}(X_2 \times Y_2)
\end{array}
$$

and

$$
\begin{array}{ccc}
H^j(X_2) \otimes H^k(Y_2) & \longrightarrow & H^{j+k}(X_2 \times Y_2) \\
\alpha^* \otimes \beta^* \downarrow & & \downarrow (\alpha \times \beta)^* \\
H^j(X_1) \otimes H^k(Y_1) & \longrightarrow & H^{j+k}(X_1 \times Y_1)
\end{array}
$$

where in all cases the horizontal maps are cross products.

Proof. This follows directly from the covariance/contravariance of the maps on homology/cohomology and the naturality of the Eilenberg-Zilber maps.   □

Next we have some algebraic properties.

Theorem 5.6.6. (1) *The cross product on both homology and cohomology is bilinear, i.e.,* $u \times (c_1 v_1 + c_2 v_2) = c_1(u \times v_1) + c_2(u \times v_2)$ *and* $(c_1 u_1 + c_2 u_2) \times v = c_1(u_1 \times v) + c_2(u_2 \times v)$ *as (co)homology classes in* $H_*(X \times Y)$ *or* $H^*(X \times Y)$.
(2) *The cross product is associative, i.e.,* $u \times (v \times w) = (u \times v) \times w$ *as (co)homology classes in* $H_*(X \times Y \times Z)$ *or* $H^*(X \times Y \times Z)$.
(3) *Let* $t : X \times Y \to Y \times X$ *by* $t(x,y) = (y,z)$. *Let* $u \in H_j(X)$ *and* $v \in H_k(Y)$. *Then* $t_*(u \times v) = (-1)^{jk}(v \times u)$. *Similarly, if* $u \in H^j(X)$ *and* $v \in H^k(Y)$, *then* $t^*(u \times v) = (-1)^{jk}(v \times u)$.

Closely related to the cross product is the slant product. We have already seen that the map

$$C^j(X) \otimes C_j(X) \longrightarrow R$$

given by evaluation of cochains on chains,

$$f \otimes \Phi \longmapsto f(\Phi)$$

gives a map on homology/cohomology

$$H^j(X) \otimes H_j(X) \longrightarrow R.$$

(This is the evaluation map of Lemma 5.5.11 that gives the map $H^j(X) \to \mathrm{Hom}\,(H_j(X), R)$ in the universal coefficient theorem, Theorem 5.5.19.)

Instead we consider the map

$$C^j(X) \otimes (C_j(X) \otimes C_k(Y)) \longrightarrow C_k(Y)$$

given by

$$f \otimes (\Phi \otimes \Psi) \longmapsto f(\Phi)\Psi.$$

We can easily extend this to a map

$$C^j(X) \otimes (C_*(X) \otimes C_*(Y))_{j+k} \longrightarrow C_k(Y)$$

by requiring

$$f \otimes (\Phi \otimes \Psi) \longmapsto 0$$

if $\Phi \in C_p(X)$, $\Psi \in C_q(Y)$, for $p \neq j$, $q \neq k$.

Again we have the Eilenberg-Zilber map

$$F : C_*(X \times Y) \longrightarrow C_*(X) \otimes C_*(Y)$$

and taking the composite gives us the slant product

$$\backslash : C^j(X) \otimes C_{j+k}(X \times Y) \longrightarrow C_k(Y).$$

Lemma 5.6.7. *The slant product induces a map*

$$\backslash : H^j(X) \otimes H_{j+k}(X \times Y) \longrightarrow H_k(Y).$$

The naturality of the slant product is more complicated to state, because of the mixed variance.

Theorem 5.6.8. *Let $\alpha : X_1 \to X_2$ and $\beta : Y_1 \to Y_2$ be maps. Let $u \in H^j(X_2)$ and $v \in H_{j+k}(X_1 \times Y_1)$. Then*

$$\beta_*(\alpha^*(u)\backslash v) = u\backslash(\alpha \times \beta)_*(v) \in H_k(Y_2).$$

Bilinearity, on the other hand, is much the same.

Theorem 5.6.9. *The slant product is bilinear, i.e., $(c_1 u_1 + c_2 u_2)\backslash v = c_1(u_1\backslash v) + c_2(u_2\backslash v)$ and $u\backslash(c_1 v_1 + c_2 v_2) = c_1(u\backslash v_1) + c_2(u\backslash v_2)$ as homology classes in Y.*

It is now easy to define the cup product and the cap product.

Definition 5.6.10. The *cup product*

$$\cup : H^j(X) \otimes H^k(X) \longrightarrow H^{j+k}(X)$$

is the map defined as follows: For $x \in H^j(X)$ and $y \in H^k(X)$,

$$x \cup y = \triangle^*(x \times y)$$

where $x \times y$ is the cross product of x and y, an element of $H^{j+k}(X \times X)$, and $\triangle^* : H^{j+k}(X \times X) \to H^{j+k}(X)$ is the map induced on cohomology by the diagonal map

$$\triangle : X \longrightarrow X \times X$$

given by $\triangle(p) = (p, p)$. ◇

Definition 5.6.11. The *cap product*

$$\cap : H^j(X) \otimes H_{j+k}(X) \longrightarrow H_k(X)$$

is the map defined as follows: For $x \in H^j(X)$ and $y \in H_{j+k}(X)$,

$$x \cap y = x\backslash\triangle_*(y)$$

where $\triangle_* : H_{j+k}(X) \to H_{j+k}(X \times X)$ is the map induced on homology by the diagonal map $\triangle$. ◇

We have defined the cup and cap products from the cross and slant products. But in fact we can recover cross and slant products from a knowledge of cup and cap products.

Lemma 5.6.12. *Let $\pi_1 : X \times Y \to X$ and $\pi_2 : X \times Y \to Y$ be projection on the first and second factors respectively.*

(1) *Let $\alpha \in H^j(X)$ and $\beta \in H^k(Y)$. Then*

$$\alpha \times \beta = \pi_1^*(\alpha) \cup \pi_2^*(\beta) = (\alpha \times 1^Y) \cup (1^X \times \beta).$$

(2) *Let $\alpha \in H^j(X)$ and $\beta \in H_{j+k}(Y)$. Then*

$$\alpha \backslash \beta = \pi_{1*}(\pi_2^*(\alpha) \cap \beta).$$

Proof. We prove the first of these. The last equality is just Lemma 5.5.23. To prove the first, let $\triangle : X \times Y \to (X \times Y) \times (X \times Y)$ be the diagonal. Then, by definition,

$$\pi_1^*(\alpha) \cup \pi_2^*(\beta) = \triangle^*(\pi_1^*(\alpha) \times \pi_2^*(\beta))$$
$$= \triangle^*(\pi_1^* \times \pi_2^*)(\alpha \times \beta)$$
$$= ((\pi_1 \times \pi_2)\triangle)^*(\alpha \times \beta)$$
$$= \alpha \times \beta$$

as $(\pi_1 \times \pi_2)\triangle$ is the identity map.   □

We now summarize some properties of cup and cap products. These follow fairly directly from our previous work and from the properties of cross and slant products.

Theorem 5.6.13. (1) *The cup product is associative.*

(2) *Cup and cap products are bilinear with respect to addition of cohomology and homology classes.*

(3) *If $\alpha \in H^j(X)$ and $\beta \in H^k(X)$,*

$$\beta \cup \alpha = (-1)^{jk} \alpha \cup \beta.$$

(4) *For any $\alpha \in H^k(X)$, $1^X \cup \alpha = \alpha$, and for any $\gamma \in H_{j+k}(X)$, $1^X \cap \gamma = \gamma$.*

(5) *For any $\alpha \in H^j(X)$, $\beta \in H^k(X)$, and $\gamma \in H_l(X)$,*

$$\alpha \cap (\beta \cap \gamma) = (\alpha \cup \beta) \cap \gamma.$$

(6) *For any $\alpha \in H^j(X)$ and $\gamma \in H_j(X)$,*

$$\varepsilon_*(\alpha \cap \gamma) = e(\alpha, \gamma).$$

(7) *For any $\alpha \in H^j(X)$, $\beta \in H^k(X)$, and $\gamma \in H_{j+k}(X)$,*

$$e(\alpha, \beta \cap \gamma) = e(\alpha \cup \beta, \gamma).$$

(8) *Let $f : X \to Y$. For any $\zeta \in H^j(Y)$ and $\alpha \in H_{j+k}(X)$,*

$$f_*(f^*(\zeta) \cap \alpha) = \zeta \cap f_*(\alpha) \in H_k(Y).$$

We also record the following more complicated relationship between cup, cap, and cross products, which includes Lemma 5.6.12(1) as a very special case.

Theorem 5.6.14. (1) *Let* $\alpha \in H^j(X)$, $\beta \in H^k(X)$, $\gamma \in H^l(Y)$, *and* $\delta \in H^m(Y)$. *Then, if* $n = j + k + l + m$,

$$(\alpha \cup \beta) \times (\gamma \cap \delta) = (-1)^{kl}(\alpha \times \gamma) \cup (\beta \times \delta) \in H^n(X \times Y).$$

(2) *Let* $\alpha \in H^j(X)$, $\beta \in H_{j+k}(X)$, $\gamma \in H^l(Y)$, *and* $\delta \in H_{l+m}(Y)$. *Then, if* $n = k + m$,

$$(\alpha \cap \beta) \times (\gamma \cap \delta) = (-1)^{(j+k)l}(\alpha \times \gamma) \cap (\beta \times \delta) \in H_n(X \times Y).$$

This follows from the previous properties we have obtained with enough careful attention to detail (including signs). Note in (2) that the first and third cross products are in homology and the second is in cohomology.

Many of the properties of the cup and cap product that we have stated can be subsumed under the following theorem. (See Definition A.1.13 for the definition of a graded algebra and module.)

Theorem 5.6.15. (1) *For any nonempty space* X, $\mathscr{S} = \bigoplus_i H^i(X)$ *is a graded commutative R-algebra and* $\mathscr{N} = \bigoplus_i H_i(X)$ *is a left* $\mathscr{S}$-*module.*
(2) *Let* X *and* Y *be nonempty spaces and let* $f : X \to Y$ *be a map. Let* $\mathscr{T} = \bigoplus_i H^i(X)$ *and* $\mathscr{S} = \bigoplus_i H^i(Y)$. *Then* $f^* : \mathscr{S} \to \mathscr{T}$ *is an R-algebra homomorphism.*

Corollary 5.6.16. *If* X *and* Y *are homotopy equivalent, then* $\bigoplus_i H^i(X)$ *and* $\bigoplus_i H^i(Y)$ *are isomorphic as graded R-algebras.*

We now consider homology and cohomology of pairs. Again we can define cup and cap products (through with some mild restrictions) and the results we obtain are almost the same.

Theorem 5.6.17. *Let* X *be a space and let* C *and* D *be subspaces of* X. *Assume that* $\{X \times C, D \times X\}$ *and* $\{C, D\}$ *are both excisive couples. Then there is a cup product*

$$\cup : H^j(X, C) \otimes H^k(X, D) \longrightarrow H^{j+k}(X, C \cup D)$$

and a cap product

$$\cap : H^j(X, C) \otimes H_{j+k}(X, C \cup D) \longrightarrow H_k(X, D).$$

Proof. The condition that $\{X \times C, D \times X\}$ be excisive is necessary in order to apply the Eilenberg-Zilber theorem, and the condition that $\{C, D\}$ be excisive is necessary to obtain the analog of Lemma 5.6.12. Otherwise, the constructions are entirely analogous (though more complicated).   □

The following particulary useful special cases of this theorem are worth pointing out.

Corollary 5.6.18. *Let (X,A) be a pair. In part (1), assume that $\{X \times A, A \times X\}$ is an excisive couple.*

(1) *There is a cup product*

$$\cup : H^j(X,A) \otimes H^k(X,A) \longrightarrow H^{j+k}(X,A)$$

and a cap product

$$\cap : H^j(X,A) \otimes H_{j+k}(X,A) \longrightarrow H_k(X,A).$$

(2) *There is a cup product*

$$\cup : H^j(X) \otimes H^k(X,A) \longrightarrow H^{j+k}(X,A)$$

and a cap product

$$\cap : H^j(X) \otimes H_{j+k}(X,A) \longrightarrow H_k(X,A).$$

(3) *There is a cup product*

$$\cup : H^j(X,A) \otimes H^k(X) \longrightarrow H^{j+k}(X,A)$$

and a cap product

$$\cap : H^j(X,A) \otimes H_{j+k}(X,A) \longrightarrow H_k(X).$$

Proof. (1) This is the special case $C = D = A$ of Theorem 5.6.17.
(2) This is the special case $C = A$, $D = \emptyset$ of Theorem 5.6.17.
(3) This is the special case $C = \emptyset$, $D = A$ of Theorem 5.6.17.

In (1), we need the hypothesis that $\{X \times A, A \times X\}$ is excisive. But in these special cases, all the other excisiveness hypotheses are automatic. $\square$

Remark 5.6.19. Note that if A is nonempty then $\bigoplus_i H^i(X,A)$ is a graded commutative ring *without* 1. For example, if X is a path connected space and A is a nonempty subspaces of X, then $H^0(X,A) = \{0\}$. $\Diamond$

Otherwise the analogous statements all hold and we will not bother to explicitly formulate them.

5.7 Some Applications of the Cup Product

In this section we give several applications of the cup product. We begin by considering a pair of spaces that have the same (co)homology groups, and show they have different cohomology ring structures, so they cannot be homotopy equivalent. Then we compute the ring structure on the cohomology of $\mathbb{R}P^n$ and $\mathbb{C}P^n$, and use the result for $\mathbb{R}P^n$ to derive the Borsuk-Ulam theorem.

Example 5.7.1. We take $\mathbb{Z}$ coefficients. Let $p, q \geq 1$. Then $H^p(S^p) \cong \mathbb{Z}$ and we choose a generator α. Also, $H^q(S^q) \cong \mathbb{Z}$ and we choose a generator β. Now consider $H^*(S^p \times S^q)$. It is given, in case $p \neq q$, by

$$H^n(S^p \times S^q) = \begin{cases} \mathbb{Z} & n = p+q \\ \mathbb{Z} & n = p, q \\ \mathbb{Z} & n = 0 \\ 0 & \text{otherwise} \end{cases}$$

and in case $p = q$ by

$$H^n(S^p \times S^q) = \begin{cases} \mathbb{Z} & n = p+q \\ \mathbb{Z} \oplus \mathbb{Z} & n = p = q \\ \mathbb{Z} & n = 0 \\ 0 & \text{otherwise.} \end{cases}$$

If $\pi_1 : S^p \times S^q \to S^p$ is projection on the first factor and $\pi_2 : S^p \times S^q \to S^q$ is projection on the second factor, then, by Lemma 5.5.23, in case $p \neq q$, $H^p(S^p \times S^q)$ is generated by $\tilde{\alpha} = \pi_1^*(\alpha) \otimes 1$ and $H^q(S^p \times S^q)$ is generated by $\tilde{\beta} = 1 \otimes \pi_2^*(\beta)$, while in case $p = q$, $H^p(S^p \times S^q)$ is generated by the two classes $\tilde{\alpha}$ and $\tilde{\beta}$. In either case, by the Künneth formula, we have an isomorphism $H^p(S^p) \otimes H^q(S^q) \to H^{p+q}(S^p \times S^q)$ which is given by the cross-product. Thus $H^{p+q}(S^p \times S^q)$ is generated by $\tilde{\gamma} = \alpha \times \beta$. But by Lemma 5.6.12 this gives

$$\tilde{\gamma} = \tilde{\alpha} \cup \tilde{\beta}. \qquad \diamond$$

Example 5.7.2. Again we take $\mathbb{Z}$ coefficients. Let $p, q \geq 1$. Let $Y = S^p \vee S^q \vee S^{p+q}$, i.e., the union of S^p, S^q, and S^{p+q} with all three spaces identified at one point.

Let $Z = S^p \vee S^q \subset Y$ and note that we have a retraction $f : Y \to Z$ given by collapsing S^{p+q} to the identification point. Then $f^* : H^n(Z) \to H^n(Y)$ is an isomorphism for $n = p, q$. Let α be a generator of $H^p(S^p)$ and β be a generator of $H^q(S^q)$. Let $\tilde{\alpha} = f^*(\alpha)$ and $\tilde{\beta} = f^*(\beta)$. Note that $\alpha \cup \beta = 0$ as $\alpha \cup \beta \in H^{p+q}(Z) = \{0\}$. But then

$$\tilde{\alpha} \cup \tilde{\beta} = f^*(\alpha) \cup f^*(\beta) = f^*(\alpha \cup \beta) = f^*(0) = 0. \qquad \diamond$$

Theorem 5.7.3. *Let* $X = S^p \times S^q$ *and* $Y = S^p \vee S^q \vee S^{p+q}$. *If p and q are not both 0, then X and Y are not homotopy equivalent.*

Proof. If $p = 0$ or $q = 0$ this is trivial.

Suppose $p, q \geq 1$. Then by Examples 5.7.1 and 5.7.2 X and Y have nonisomorphic cohomology rings, so by Corollary 5.6.16 they are not homotopy equivalent.   □

Here is another computation of a cohomology ring.

Lemma 5.7.4. (1) *Let* $n \geq 1$. *Let* $\alpha \in H^1(\mathbb{R}P^n; \mathbb{Z}_2)$ *be a generator. Then* $\alpha^k \in H^k(\mathbb{R}P^n; \mathbb{Z}_2)$ *is a generator for all* $k \leq n$.
(2) *Let* $n \geq 1$. *Let* $\alpha \in H^2(\mathbb{C}P^n; \mathbb{Z}_2)$ *be a generator. Then* $\alpha^k \in H^{2k}(\mathbb{C}P^n; \mathbb{Z})$ *is a generator for all* $k \leq n$.

Proof. The proofs are almost identical so we prove (1).

We prove this by induction on n. The case $n = 1$ is trivial. Assume the lemma is true for n and consider $\mathbb{R}P^{n+1}$. The inclusion $\mathbb{R}P^n \hookrightarrow \mathbb{R}P^{n+1}$ induces isomorphisms on (co)homology in dimensions at most n. That readily implies that $\alpha^k \in H^k(\mathbb{R}P^{n+1}; \mathbb{Z}_2)$ is a generator for all $k \leq n$, so it remains to prove that α^{n+1} generates $H^{n+1}(\mathbb{R}P^{n+1}; \mathbb{Z}_2)$.

Let $\mathbb{R}P^n$ have homogeneous coordinates $[x_0, \ldots, x_n]$ and let $\mathbb{R}P^1$ have homogeneous coordinates $[y_0, y_1]$. Then we have a map $r : \mathbb{R}P^n \times \mathbb{R}P^1 \to \mathbb{R}P^{2n+1}$ given by

$$([x_0, \ldots, x_n], [y_0, y_1]) \longmapsto [x_0 y_0, \ldots, x_n y_0, x_0 y_1, \ldots, x_n y_1].$$

Let $\alpha \in H^1(\mathbb{R}P^{2n+1}; \mathbb{Z}_2)$ be the generator. We will show that $\alpha^{n+1} \in H^{n+1}(\mathbb{R}P^{2n+1}; \mathbb{Z}_2)$ is the generator. Since the inclusion $\mathbb{R}P^{n+1} \hookrightarrow \mathbb{R}P^{2n+1}$ induces isomorphisms on homology in dimensions at most $n + 1$, that yields the inductive step.

Consider the map $p : \mathbb{R}P^n \to \mathbb{R}P^n \times \{[1, 0]\} \hookrightarrow \mathbb{R}P^n \times \mathbb{R}P^1 \to \mathbb{R}P^{2n+1}$. Then $\beta = p^*(\alpha)$ is the generator of $H^1(\mathbb{R}P^n; \mathbb{Z}_2)$ and by the inductive hypothesis $\beta^n = (p^*(\alpha))^n = p^*(\alpha^n)$ is the generator of $H^n(\mathbb{R}P^n; \mathbb{Z}_2)$. Similarly, if $q : \mathbb{R}P^1 \to \{[1, \ldots, 0]\} \times \mathbb{R}P^1 \hookrightarrow \mathbb{R}P^n \times \mathbb{R}P^1 \to \mathbb{R}P^{2n+1}$, $\gamma = q^*(\alpha)$ is the generator of $H^1(\mathbb{R}P^1; \mathbb{Z}_2)$.

But then, by Lemma 5.5.23 $\pi_1^*(\beta^n) \times 1$ is the generator of $H^n(\mathbb{R}P^n; \mathbb{Z}_2) \otimes H^0(\mathbb{R}P^n; \mathbb{Z}_2) \subseteq H^n(\mathbb{R}P^n \times \mathbb{R}P^1; \mathbb{Z}_2)$ and $1 \times \pi_2^*(\gamma)$ is the generator of $H^0(\mathbb{R}P^n; \mathbb{Z}_2) \otimes H^1(\mathbb{R}P^1; \mathbb{Z}_2) \subseteq H^n(\mathbb{R}P^n \times \mathbb{R}P^1; \mathbb{Z}_2)$. But then by the Künneth formula

$$(\pi_1^*(\beta^n) \times 1) \cup (1 \times \pi_2^*(\gamma)) = \pi_1^*(\beta^n) \times \pi_2^*(\gamma)$$

generates $H^{n+1}(\mathbb{R}P^n \times \mathbb{R}P^1; \mathbb{Z}_2)$.

But this class is just

$$r^*(\alpha^n) \cup r^*(\alpha) = r^*(\alpha)^n \cup r^*(\alpha) = r^*(\alpha^n \cup \alpha) = r^*(\alpha^{n+1})$$

so α^{n+1} must be the generator of $H^{n+1}(\mathbb{R}P^{2n+1};\mathbb{Z}_2)$. $\qquad\qquad\qquad\square$

Corollary 5.7.5. *Let $n > m \geq 1$. If $f : \mathbb{R}P^n \to \mathbb{R}P^m$ is any map, then $f_* : H_1(\mathbb{R}P^n) \to H_1(\mathbb{R}P^m)$ is the zero map.*

Proof. If $m = 1$, then $H_1(\mathbb{R}P^n) = \mathbb{Z}_2$ and $H_1(\mathbb{R}P^m) = \mathbb{Z}$ and the only map from $\mathbb{Z}_2$ to $\mathbb{Z}$ is the zero map. Suppose $m > 1$. Then $H_1(\mathbb{R}P^n) = \mathbb{Z}_2$ and $H_1(\mathbb{R}P^m) = \mathbb{Z}_2$, so f_* is either an isomorphism or the zero map.

Assume f_* is an isomorphism. Then $f_* : H_1(\mathbb{R}P^n;\mathbb{Z}_2) \to H_1(\mathbb{R}P^m;\mathbb{Z}_2)$ is an isomorphism, by the universal coefficient theorem, and hence so is $f^* : H^1(\mathbb{R}P^m;\mathbb{Z}_2) \to H^1(\mathbb{R}P^n;\mathbb{Z}_2)$. Let α be the generator of $H^1(\mathbb{R}P^m;\mathbb{Z}_2)$ so that $\beta = f^*(\alpha)$ is the generator of $H^1(\mathbb{R}P^n;\mathbb{Z}_2)$. Then by Lemma 5.7.4,

$$0 \neq \beta^n = (f^*(\alpha))^n = f^*(\alpha^n) = f^*(0) = 0,$$

a contradiction. (Note $\alpha^n = 0$ as $\alpha^n \in H^n(\mathbb{R}P^m;\mathbb{Z}_2) = 0$ as $n > m$.) $\qquad\square$

As a consequence of this we have the famous Borsuk-Ulam theorem.

Theorem 5.7.6 (Borsuk-Ulam).   *Let $n \geq 1$. Let $f : S^n \to \mathbb{R}^n$ be any map. Then there is an $x \in S^n$ with $f(x) = f(-x)$.*

Proof. Assume there is no such point x and define $g : S^n \to S^{n-1}$ by

$$g(x) = \frac{f(x) - f(-x)}{|f(x) - f(-x)|}.$$

Observe that $g(-x) = -g(x)$.

Recall that for any n, $\mathbb{R}P^n$ is the quotient of S^n by the antipodal map $a(x) = -x$. Then we have the 2-fold covering maps $p_k : S^k \to \mathbb{R}P^k$, $k = n-1$ or n, and a well-defined map $h : \mathbb{R}P^n \to \mathbb{R}P^{n-1}$ given by

$$h(p_n(x)) = p_{n-1}(g(x)).$$

Then we have a commutative diagram

$$
\begin{array}{ccc}
S^n & \xrightarrow{\ g\ } & S^{n-1} \\[4pt]
{\scriptstyle p_n}\downarrow & & \downarrow{\scriptstyle p_{n-1}} \\[4pt]
\mathbb{R}P^n & \xrightarrow{\ h\ } & \mathbb{R}P^{n-1}.
\end{array}
$$

By Theorem 2.2.8, the map h lifts to a map $\tilde{h} : \mathbb{R}P^n \to S^{n-1}$ if and only if

$$h_*(\pi_1(\mathbb{R}P^n)) \subseteq (p_{n-1})_*(\pi_1(S^{n-1})).$$

We note that $\pi_1(\mathbb{R}P^n)$ and $\pi_1(\mathbb{R}P^{n-1})$ are abelian, so they are isomorphic to $H_1(\mathbb{R}P^n)$ and $H_1(\mathbb{R}P^{n-1})$ respectively, and thus that will be the case if and only if

$$h_*(H_1(\mathbb{R}P^n)) \subseteq (p_{n-1})_*(H_1(S^{n-1})).$$

But by Corollary 5.7.5, h_* is the 0 map, so this is certainly true.

Now to complete the proof. We observe that $p_{n-1}\tilde{h}p_n = p_{n-1}g$, i.e., $\tilde{h}p_n$ and g are both liftings of h to a map from S^n to S^{n-1}.

Pick $x_0 \in S^n$. Then $\tilde{h}p_n(x_0)$ is one of the two points in S^{n-1} covering $p_{n-1}(g(x_0))$, i.e.,

$$\tilde{h}p_n(x_0) = g(x_0) \quad \text{or} \quad -g(x_0) = g(-x_0).$$

If $\tilde{h}p_n(x_0) = g(x_0)$, set $x_1 = x_0$. Otherwise set $x_1 = -x_0$ and observe

$$\tilde{h}p_n(x_1) = \tilde{h}p_n(-x_0) = \tilde{h}p_n(x_0) = g(-x_0) = g(x_1).$$

Thus in any case we have found a point x_1 on which both lifts of h agree. By the uniqueness part of Theorem 2.2.8, this implies that they agree everywhere. But that is a contradiction as on the one hand, since the lifts agree at $-x_1$,

$$\tilde{h}p_n(-x_1) = g(-x_1),$$

and on the other hand, since $p_n(-x_1) = p_n(x_1)$,

$$\tilde{h}p_n(-x_1) = \tilde{h}p_n(x_1) = g(x_1),$$

but

$$g(-x_1) = -g(x_1) \neq g(x_1).$$

$\square$

5.8   Exercises

Exercise 5.8.1. Let A be a nonempty subspace of X. Show there is an exact homology sequence

$$\cdots \longrightarrow H_i(A) \longrightarrow H_i(X) \longrightarrow H_i(X,A) \longrightarrow \cdots$$

$$H_1(X,A) \longrightarrow \tilde{H}_0(A) \longrightarrow \tilde{H}_0(X) \longrightarrow H_0(X,A) \longrightarrow 0.$$

Exercise 5.8.2.   (a)   Let $\alpha \in H_k(X,A)$. Show there is a compact pair $(Y,B) \subseteq (X,A)$ and an element $\beta \in H_k(X,A)$ such that $\alpha = i_*(\beta)$, where $i : (Y,B) \to (X,A)$ is the inclusion.

(b) Let $(Z,C) \subseteq (X,A)$ be a compact pair and let $i : (Z,C) \to (X,A)$ be the inclusion. Let $\gamma \in H_k(Z,C)$ with $i_*(\gamma) = 0$. Show there is a compact pair (Y,B) with $(Z,C) \subseteq (Y,B) \subseteq (X,A)$ such that $j_*(\gamma) = 0$, where $j : (Z,C) \to (Y,B)$ is the inclusion.

Note this shows that singular homology is compactly supported.

Exercise 5.8.3. Find an example of a path connected space X with $H_1(X) = 0$, and two path connected closed subspaces X_1 and X_2 of X with $X = X_1 \cup X_2$, where $A = X_1 \cap X_2$ is not path-connected.

Note this shows that the Mayer-Vietoris sequence cannot be exact in this situation. In fact, give an example of this where X, X_1, and X_2 are each contractible, and where A is the union of two contractible path components.

Exercise 5.8.4. Compute $H_*(\mathbb{R}P^n \times \mathbb{R}P^m; G)$, where (a) $G = \mathbb{Z}_2$ or (b) $G = \mathbb{Z}$.

Exercise 5.8.5. Prove Theorem 5.5.20: The Euler characteristic of any space with finitely generated homology may be computed using coefficients the integers $\mathbb{Z}$ or any field, and using homology or cohomology (i.e., that the answers so obtained are the same in all cases).

Exercise 5.8.6. Let X and Y be any spaces with finitely generated homology. Show that $\chi(X \times Y) = \chi(X)\chi(Y)$.

Exercise 5.8.7. Let (X,A) be any pair such that X and A both have finitely generated homology. Assume A is nonempty. Show that $\chi(X/A) = \chi(X) - \chi(A) + 1$.

Exercise 5.8.8. Let $\mathbb{F}$ be a field. Describe the ring structure on $H^*(X \times Y; \mathbb{F})$ in terms of the ring structures on $H^*(X; \mathbb{F})$ and $H^*(Y; \mathbb{F})$.

Exercise 5.8.9. The join $X * Y$ of two spaces X and Y is the quotient space $X \times [-1,1] \times Y/ \sim$ where $\sim$ is the identification $(x, -1, y_1) \sim (x, -1, y_2)$ for any $y_1, y_2 \in Y$ and $(x_1, 1, y) \sim (x_2, 1, y)$ for any $x_1, x_2 \in X$.

(a) Show that $S^0 * Y$ is homeomorphic to ΣY, the suspension of Y.
(b) Show that $S^i * S^j$ is homeomorphic to S^{i+j+1}.
(c) Show that $S^k * Y$ is homeomorphic to $\Sigma^{k+1}Y$, the $(k+1)$-fold suspension $\Sigma(\Sigma(\cdots(\Sigma Y)))$ of Y.

Exercise 5.8.10. Let X and Y be spaces with $\chi(X)$ and $\chi(Y)$ defined. Find $\chi(X * Y)$.

Exercise 5.8.11. Compute $H_*(X * Y; G)$ in terms of $H_*(X; G)$ and $H_*(Y; G)$ in each of the following cases: (a) G is a field, and (b) $G = \mathbb{Z}$ and both $H_*(X; G)$ and $H_*(Y; G)$ are torsion-free.

Exercise 5.8.12. A space has category n if it can be written as the union of n open contractible subspaces, but no fewer. A space has cup length m if there are m positive dimensional cohomology classes whose cup product is nonzero, but no more. Show that the category of a space is greater than its cup length.

Exercise 5.8.13. Show that each of $\mathbb{R}P^n$ and $\mathbb{C}P^n$ have category $n + 1$, for every n.

Chapter 6
Manifolds

Manifolds are a particularly important class of topological spaces. On the one hand, there are branches of topology entirely dedicated to studying them, and on the other hand, they appear throughout much of mathematics. It would take us too far afield to describe how they arise, but they have very special properties from the point of view of algebraic topology. It is these that we investigate here.

We begin by defining manifolds, then investigate orientations, and finally arrive at duality theorems.

Recall that a topological space is Hausdorff if any two distinct points have disjoint open neighborhoods, and separable, or second countable, if it has a countable basis for its topology.

6.1 Definition and Examples

Definition 6.1.1. A topological space M is an n-dimensional *manifold* (or *n-manifold*, for short) if M is a separable Hausdorff space and if every point $x \in M$ has a neighborhood U_x that is homeomorphic to $\mathbb{R}^n$. ◇

Lemma 6.1.2. *M is an n-manifold if and only if M is a separable Hausdorff space and M has an open cover $\{U_\alpha\}$ with each U_α homeomorphic to $\mathbb{R}^n$.*

Definition 6.1.3. Let M be an n-manifold as in Lemma 6.1.2 and for each α, let $\varphi_\alpha : \mathbb{R}^n \to U_\alpha$ be a homeomorphism. Then $\{(U_\alpha, \varphi_\alpha)\}$ is an *atlas* for M and each $(U_\alpha, \varphi_\alpha)$ is a *coordinate patch*. ◇

In this situation we will often simply refer to U_α as a coordinate patch when it is not important to specify the homeomorphism φ_α.

© Springer International Publishing Switzerland 2014
S.H. Weintraub, *Fundamentals of Algebraic Topology*, Graduate Texts
in Mathematics 270, DOI 10.1007/978-1-4939-1844-7_6

Remark 6.1.4. Since $\mathbb{R}^n$ and $\mathring{D}^n$ (the interior of the unit disk in $\mathbb{R}^n$) are homeomorphic (e.g., by the map $f(x) = x/(|x|+1)$), we may replace $\mathbb{R}^n$ by $\mathring{D}^n$ in the above.   ◇

Definition 6.1.5. M is a *compact n-manifold* if it is an n-manifold that is compact as a topological space.   ◇

Here are some examples of manifolds.

Example 6.1.6. (1) $\mathbb{R}^n$ is an n-manifold. More generally, any open subset of $\mathbb{R}^n$ is an n-manifold.
(2) S^n is a compact n-manifold. (For any point $x \in S^n$, $S^n - \{x\}$ is homeomorphic to $\mathbb{R}^n$.)
(3) If M is an m-manifold and N is n-manifold, then $M \times N$ is an $(m+n)$-manifold.
(4) If $p : Y \to X$ is a covering projection, then X is a manifold if and only if Y is a manifold. If Y is compact, then X is compact. If X is compact and p is a finite covering, then Y is compact. Thus we see that $\mathbb{R}P^n$ is a compact n-manifold, and each lens space $L^{2m-1}(k; j_1, \ldots, j_m)$ of Example 2.3.4 is a compact $(2m-1)$-manifold.
(5) Consider $\mathbb{R}P^n = \{[x_0, \ldots, x_n]\}$. For $i = 0, \ldots, n$, let $p_i = [x_0, \ldots, x_n]$ where $x_i = 1$ and $x_j = 0$ for $j \neq i$. Then $\mathbb{R}P^n - \{p_i\}$ is homeomorphic to $\mathbb{R}^n$, again showing that $\mathbb{R}P^n$ is an n-manifold. Next consider $\mathbb{C}P^n = \{[z_0, \ldots, z_n]\}$ and similarly let $q_i = [z_0, \ldots, z_n]$ where $z_i = 1$ and $z_j = 0$ for $j \neq i$. Then $\mathbb{C}P^n - \{q_i\}$ is homeomorphic to $\mathbb{C}^n$, so $\mathbb{C}P^n$ is a $(2n)$-manifold. Furthermore, $\mathbb{R}P^n$ is the image of S^n under the map $(x_0, \ldots, x_n) \mapsto [x_0, \ldots, x_n]$, and $\mathbb{C}P^n$ is the image of S^{2n+1} under the map $(z_0, \ldots, z_n) \mapsto [z_0, \ldots, z_n]$, so $\mathbb{R}P^n$ and $\mathbb{C}P^n$ are both compact.   ◇

Closely related to the notion of a manifold is that of a manifold with boundary. We let $\mathbb{R}^n_+$ denote the closed half-space in $\mathbb{R}^n$ given by $\mathbb{R}^n_+ = \{(x_1, \ldots, x_n) \in \mathbb{R}^n \mid x_n \geq 0\}$. By definition, $\mathbb{R}^0_+ = \emptyset$. For $n \geq 1$, we let $\partial \mathbb{R}^n_+ = \{(x_1, \ldots, x_n) \in \mathbb{R}^n \mid x_n = 0\}$.

Definition 6.1.7. A topological space M is an *n-dimensional manifold with boundary* (or *n-manifold with boundary*, for short) if M is a separable Hausdorff space and if every point $x \in M$ has a neighborhood U_x that is homeomorphic to $\mathbb{R}^n$ or $\mathbb{R}^n_+$.

The *interior* $\text{int}(M) = \{x \in M \mid U_x \text{ is homeomorphic to } \mathbb{R}^n\}$ and the *boundary* $\partial M = \{x \in M \mid U_x \text{ is homeomorphic to } \mathbb{R}^n_+\}$.   ◇

Theorem 6.1.8. *If M is an n-manifold with nonempty boundary then $\text{int}(M)$ is an n-manifold and ∂M is an $(n-1)$-manifold.*

Proof. The first statement is clear. As for the second, if $x \in \partial M$ and $\varphi_x : \mathbb{R}^n_+ \to U$ is a homeomorphism with $x \in \varphi_x(\partial \mathbb{R}^n_+)$, then $\varphi_x|\partial \mathbb{R}^n_+$ is a homeomorphism of $\partial \mathbb{R}^n_+$ (itself homeomorphic to $\mathbb{R}^{n-1}$) to the neighborhood $U \cap \partial M$ of x in ∂M.   □

We have the analogs of Lemma 6.1.2 and Definition 6.1.3.

Lemma 6.1.9. *M is an n-manifold with boundary if and only if M is a Hausdorff space and M has an open cover $\{U_\alpha\}$ with each U_α homeomorphic to either $\mathbb{R}^n$ or $\mathbb{R}^n_+$.*

Definition 6.1.10. Let M be an n-manifold with boundary as in Lemma 6.1.9 and for each α, let $\varphi_\alpha : \mathbb{R}^n \to U_\alpha$ or $\varphi_\alpha : \mathbb{R}^n_+ \to U_\alpha$ be a homeomorphism. Then $\{(U_\alpha, \varphi_\alpha)\}$ is an *atlas* for M and each $(U_\alpha, \varphi_\alpha)$ is a *coordinate patch*. ◇

Example 6.1.11. (1) Every manifold M is a manifold with empty boundary.
(2) For every $n \geq 1$, D^n is a manifold with boundary whose boundary is S^{n-1}.
(3) Let M be an arbitrary n-manifold, $n \geq 1$. Let $\varphi_\alpha : \mathbb{R}^n \to U_\alpha \subseteq M$ be a coordinate patch. Let $\mathring{D}^n$ be the open unit ball in $\mathbb{R}^n$. Then $M - \varphi_\alpha(\mathring{D}^n)$ is a manifold with boundary whose boundary $\varphi_\alpha(S^{n-1})$ is homeomorphic to S^{n-1}. ◇

We have the following very important (and highly nontrivial) homological properties of manifolds.

Theorem 6.1.12. (1) *Let M be an n-manifold (possibly with boundary). Then $H_k(M) = H^k(M) = 0$ for all $k > n$.*
(2) *Let M be a compact n-manifold (possibly with boundary). Then $H_k(M)$ and $H^k(M)$ are finitely generated for all k.*

6.2 Orientations

In this section we develop the notion of orientation. We let $G = \mathbb{Z}/2\mathbb{Z}$ or $\mathbb{Z}$.

Lemma 6.2.1. *Let M be an n-manifold and let $x \in M$ be arbitrary. Then $H_n(M, M - x; G)$ is isomorphic to G.*

Proof. Let $(U_\alpha, \varphi_\alpha)$ be a coordinate patch with $x \in \varphi_\alpha$. Let $p = \varphi_\alpha^{-1}(x)$, $p \in \mathbb{R}^n$. Then we have maps

$$(\mathbb{R}^n, \mathbb{R}^n - \{p\}) \longrightarrow (U_\alpha, U_\alpha - \{x\}) \longrightarrow (M, M - x)$$

where the first map is φ_α and the second map is the inclusion. Now $H_n(\mathbb{R}^n, \mathbb{R}^n - \{p\}; G)$ is isomorphic to G. The first map induces an isomorphism on homology as it is a homeomorphism of pairs and the second map is an isomorphism as it is excisive. $(\overline{M - U_\alpha} = M - U_\alpha \subset M - x = \text{int}(M - x))$. Thus $H_n(M, M - x; G)$ is isomorphic to G as well. □

Definition 6.2.2. A *local G-orientation* on M at x is a choice of isomorphism $\overline{\varphi}_x : G \to H_n(M, M - x; G)$. ◇

To (attempt to) define a G-orientation on M we need to see how local G-orientations fit together.

Definition 6.2.3. (1) Let x and y both lie in some coordinate patch U_α. Let $\varphi_\alpha :$ $\mathbb{R}^n \to U_\alpha$ and set $p = \varphi_\alpha^{-1}(x)$ and $q = \varphi_\alpha^{-1}(y)$. Let D be a closed disc in $\mathbb{R}^n$ containing both p and q. The local G-orientations $\overline{\varphi}_x$ and $\overline{\varphi}_y$ are *compatible* if the following diagram commutes

$$
\begin{array}{ccccc}
H_n(M, M - x; G) & \xleftarrow{\cong} & H_n(U_\alpha, U_\alpha - x; G) & \xleftarrow{(\varphi_\alpha)_*} & H_n(\mathbb{R}^n, \mathbb{R}^n - p; G) \\
\uparrow{\overline{\varphi}_x} & & & & \uparrow{\cong} \\
G & & & & H_n(\mathbb{R}^n, \mathbb{R}^n - D; G) \\
\downarrow{\overline{\varphi}_y} & & & & \downarrow{\cong} \\
H_n(M, M - y; G) & \xleftarrow{\cong} & H_n(U_\alpha, U_\alpha - y; G) & \xleftarrow{(\varphi_\alpha)_*} & H_n(\mathbb{R}^n, \mathbb{R}^n - q; G)
\end{array}
$$

where the unlabelled maps are all induced by inclusions. $\Diamond$

(2) Let x and y both lie in the same component of M. Then $\overline{\varphi}_x$ and $\overline{\varphi}_y$ are *compatible* if there is a sequence of points $x_0 = x, x_1, \ldots, x_k = y$ with x_i and x_{i+1} both lying in some coordinate patch, for each i, and φ_{x_i} and $\varphi_{x_{i+1}}$ are compatible for each i. $\Diamond$

Remark 6.2.4. There is always such a sequence of points as we may let $f : I \to M$ be an arbitrary map with $f(0) = x$ and $f(y) = 1$. Then $f(I)$ is a compact set so is covered by finitely many coordinate patches, and then $x_0, \ldots, x_k$ are easy to find. Thus the condition in the definition is the condition on the local G-orientations. It is also easy to check that this condition is independent of the choice of intermediate points $x_1, \ldots, x_{k-1}$. $\Diamond$

Definition 6.2.5. The n-manifold M is *G-orientable* if there exists a compatible collection of local G-orientations $\{\overline{\varphi}_x\}$ for all points $x \in M$. In that case a choice of mutually compatible $\{\overline{\varphi}_x\}$ is a *G-orientation* of M.

If there is no compatible collection of local G-orientations on M then M is *G-nonorientable*. $\Diamond$

We have so far let $G = \mathbb{Z}/2\mathbb{Z}$ or $\mathbb{Z}$. But now we see a big difference between these two cases.

Theorem 6.2.6. *Every manifold is $\mathbb{Z}/2\mathbb{Z}$-orientable.*

Proof. Following the diagram in Definition 6.1.3 all the way around from G to G gives an isomorphism from G to G, and $\overline{\varphi}_x$ and $\overline{\varphi}_y$ are compatible if and only if this isomorphism is the identity. But the only isomorphism $\overline{\varphi} : \mathbb{Z}/2\mathbb{Z} \to \mathbb{Z}/2\mathbb{Z}$ is the identity. $\square$

Referring to the proof of this theorem, we see that in case $G = \mathbb{Z}$ we have an isomorphism $\varphi : \mathbb{Z} \to \mathbb{Z}$. But now there are two isomorphisms, the identity (i.e., multiplication by 1) and multiplication by -1. So a priori, M might or might not be orientable.

Theorem 6.2.7. (1) *Let M be the union of components $M = M_1 \cup M_2 \cup \cdots$. Then M is $\mathbb{Z}$-orientable if and only if each M_i is $\mathbb{Z}$-orientable.*

(2) *Let M be connected. If M is $\mathbb{Z}$-orientable, then M has exactly two $\mathbb{Z}$-orientations.*

(3) *If M has k components and is $\mathbb{Z}$-orientable, then M has 2^k $\mathbb{Z}$-orientations.*

Proof. The important thing to note is that if $\overline{\varphi}_x : \mathbb{Z} \to H_n(M, M - x; \mathbb{Z})$ is a local $\mathbb{Z}$-orientation, there is exactly one other local $\mathbb{Z}$-orientation at x, namely $-\overline{\varphi}_x$, where $-\overline{\varphi}_x : \mathbb{Z} \to H_n(M, M - x; \mathbb{Z})$ is the map defined by $-(\overline{\varphi}_x)(n) = -\overline{\varphi}_x(n)$, for $n \in \mathbb{Z}$.

Thus if we have a compatible system of local $\mathbb{Z}$-orientations $\{\overline{\varphi}_x\}$ on a connected manifold M, then there are exactly two compatible systems of local $\mathbb{Z}$-orientative on M, namely $\{\overline{\varphi}_x\}$ and $\{-\overline{\varphi}_x\}$. □

Having made our point, we now drop the $\mathbb{Z}$ and use standard mathematical language.

Definition 6.2.8. A *local orientation* on M is a local $\mathbb{Z}$-orientation. M is *orientable* (resp. *nonorientable*) if it is $\mathbb{Z}$-orientable (resp. $\mathbb{Z}$-nonorientable). An *orientation* of M is a $\mathbb{Z}$-orientation of M. ◇

Our first objective is to investigate when manifolds are orientable. In view of Theorem 6.2.7, we may confine our attention to connected manifolds.

Definition 6.2.9. Let M be a connected n-manifold. Let φ_x be a local orientation of M at x. Let $y \in M$ and let $f : I \to M$ with $f(0) = x$ and $f(1) = y$. The *transfer* of $\overline{\varphi}_x$ along f to y is the local orientation $f_y(\overline{\varphi}_x)$ of M at y obtained as follows: Let $0 = t_0 < t_1 < \cdots < t_k = 1$ such that, for each i, $f([t_i, t_{i+1}])$ is contained in some coordinate patch. Let $x_i = f(t_i)$ for each i, so that, in particular, $x_0 = x$ and $x_k = y$. For each i, let $\overline{\varphi}_{x_{i+1}}$ be the local orientation of M at x_{i+1} compatible with the local orientation $\overline{\varphi}_{x_i}$ of M at x_i. Then $f_y(\overline{\varphi}_x) = \overline{\varphi}_{x_k}$. ◇

The transfer of $\overline{\varphi}_x$ along f to y does not depend on the choice of points $\{t_i\}$ but it certainly may depend on the choice of f. However, we have the following result.

Theorem 6.2.10. *Let M be a connected n-manifold. Then a system of local orientations $\{\overline{\varphi}_x\}$ is an orientation of M if and only if for every $x, y \in M$ and every path $f : I \to M$ with $f(0) = x$ and $f(1) = y$, $\overline{\varphi}_y = f_y(\overline{\varphi}_x)$. In particular, M is orientable if and only if $f_y(\overline{\varphi}_x)$ is independent of the choice of f.*

Proof. If M is orientable, let $\{\overline{\varphi}_x\}$ be an orientation, i.e., a compatible system of local orientations. Then for any path f, $f_y(\overline{\varphi}_x) = \overline{\varphi}_y$ is independent of the choice of f.

Conversely, if $f_y(\overline{\varphi}_x)$ is independent of the choice of f, we may obtain a compatible system of local orientations as follows: Choose a point $x \in M$ and a local orientation $\overline{\varphi}_x$. Then for a point $y \in M$, choose any path f from x to y and let $\overline{\varphi}_y = f_y(\overline{\varphi}_x)$. □

This result makes it crucial to investigate the dependence of $f_y(\overline{\varphi}_x)$ on the choice of the path f. We do that now.

Lemma 6.2.11. (1) *Let x and y be two points in M that are both contained in some coordinate patch U_α. Then for any two paths f and g from x to y with $f(I) \subset U_\alpha$ and $g(I) \subset U_\alpha$, $f_y(\overline{\varphi}_x) = g_y(\overline{\varphi}_x)$.*
(2) *Let x and y be any two points in M. If f and g are any two paths in M that are homotopic rel $\{0,1\}$, then $f_y(\overline{\varphi}_x) = g_y(\overline{\varphi}_x)$.*

Proof. (1) Since I is compact, $f(I)$ is a compact subset of U_α, and hence $\varphi_\alpha^{-1}(f(I))$ is a compact subset of $\mathbb{R}^n$, as is $\varphi_\alpha^{-1}(g(I))$. But then we may choose the disc D in Definition 6.2.3 so large that D includes $\varphi_\alpha^{-1}(f(I)) \cup \varphi_\alpha^{-1}(g(I))$.
(2) Let $F : I \times I \to M$ be a homotopy of f to g rel $\{0,1\}$. Then $F(I \times I)$ is a compact subset of M, so is covered by finitely many coordinate patches. It is then easy to see that there is some k such that if $I \times I$ is divided into k^2 subsquares $J_{i,j} = [i/k, (i+1)/k] \times [j/k, (j+1)/k]$, each $F(J_{i,j})$ is contained in a single coordinate patch.

But then we have a sequence of paths from f to g, or, more precisely, to the constant path followed by g followed by the constant path, with each path giving the same transferred orientation, according to part (1) and the following picture:

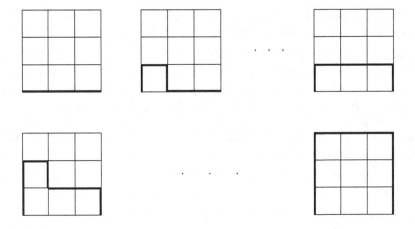

$\square$

Lemma 6.2.12. (1) *Let f be a path in M from x to y and let g be a path in M from y to z. Let $h = fg$ be the path from x to z obtained by first following f and then following g. Then $h_z(\overline{\varphi}_x) = g_z(f_y(\overline{\varphi}_x))$.*
(2) *Let f and g be paths in M from x to y and let $\overline{g}$ be the path in M from y to x which is the reverse of g. Let $h = f\overline{g}$, a path from x to x. Then $f_y(\overline{\varphi}_x) = g_y(\overline{\varphi}_x)$ if and only if $h_x(\overline{\varphi}_x) = \overline{\varphi}_x$.*

Before continuing with our general investigation of orientability, we pause to record the following very important special case.

Theorem 6.2.13. *Let M be a connected n-manifold. If M is simply connected, then M is orientable.*

Proof. By Theorem 6.2.10, we must show that $\{f_y(\overline{\varphi}_x)\}$ is independent of the choice of f. By Lemma 6.2.12(2), that will be the case if $h_x(\overline{\varphi}_x) = \overline{\varphi}_x$ for any loop h_x based at x. But by Lemma 6.2.11, $h_x(\overline{\varphi}_x)$ only depends on the homotopy class of the loop h. If M is simply connected, then h is homotopic to the constant path c at x, and certainly $c_x(\overline{\varphi}_x) = \overline{\varphi}_x$. □

Now we return to consideration of a general connected n-manifold M. Let x be a point of M.

As we have observed, the two local orientations at x are $\overline{\varphi}_x$ and $-\overline{\varphi}_x$. We now define the orientation character.

Definition 6.2.14. Let f be a closed path in M based at x. The *orientation character* $w(f) \in \mathbb{Z}/2\mathbb{Z} = \{0, 1\}$ is defined by

$$f_x(\overline{\varphi}_x) = (-1)^{w(f)}\overline{\varphi}_x.$$

◊

Theorem 6.2.15. *A connected manifold M is orientable if and only if its orientation character $w(f) = 0$ for every loop f in M.*

Proof. In light of Lemma 6.2.12, this is just a restatement of Theorem 6.2.10. □

We now derive several other maps from the orientation character, but we mostly use the same letter to denote them.

Lemma 6.2.16. *Let M be a connected manifold. The orientation character gives a homomorphism*

$$w : \pi_1(M, x) \longrightarrow \mathbb{Z}/2\mathbb{Z}$$

defined by $w(\alpha) = w(f)$ where f is a loop in M representing $\alpha \in \pi_1(M, x)$.

Proof. By Lemma 6.2.12(2), w depends only on the homotopy class of f, and by Lemma 6.2.12(1), w is a homomorphism. □

Corollary 6.2.17. *Let M be a connected nonorientable manifold. Then M has a unique 2-fold cover N that is orientable.*

Proof. N is the cover of M corresponding to the subgroup $\mathrm{Ker}(w) \subset \pi_1(M, x)$ of index 2 as in Theorem 2.2.19. □

Lemma 6.2.18. *Let M be a manifold. The orientation character gives a homomorphism*

$$w : H_1(M; \mathbb{Z}) \longrightarrow \mathbb{Z}/2\mathbb{Z}.$$

Proof. The map $w : \pi_1(M, x) \to \mathbb{Z}/2\mathbb{Z}$ is a map to an abelian group, so factors through the abelianization of $\pi_1(M, x)$. In case M is connected that is just $H_1(M; \mathbb{Z})$ by Theorem 5.2.4. In the general case, just consider each component of M separately. □

Theorem 6.2.19. *Let M be a manifold. The orientation character gives a homomorphism*

$$w : H_1(M; \mathbb{Z}/2\mathbb{Z}) \longrightarrow \mathbb{Z}/2\mathbb{Z}.$$

Proof. The map $w : H_1(M; \mathbb{Z}) \to \mathbb{Z}/2\mathbb{Z}$ factors through $H_1(M; \mathbb{Z})/2H_1(M; \mathbb{Z})$ (i.e. $w(2\alpha) = 2w(\alpha) = 0$ for any $\alpha \in H_1(M; \mathbb{Z})$), and $H_1(M; \mathbb{Z})/2H_1(M; \mathbb{Z})$ is isomorphic to $H_1(M; \mathbb{Z}) \otimes \mathbb{Z}/2\mathbb{Z}$. In turn, this group is isomorphic to $H_1(M; \mathbb{Z}/2\mathbb{Z})$ by the universal coefficient theorem, Theorem 5.3.9, as $H_0(M; \mathbb{Z}) = \mathbb{Z}$ is torsion-free. $\square$

Corollary 6.2.20. *Let M be a manifold. If $H_1(M; \mathbb{Z}/2\mathbb{Z}) = 0$, then M is orientable.*

Recall we have the universal coefficient theorem, Theorem 5.5.12. Since $H_0(M; \mathbb{Z}) = \mathbb{Z}$, that theorem gives an isomorphism

$$e : H^1(M; \mathbb{Z}/2\mathbb{Z}) \longrightarrow \operatorname{Hom}(H_1(M), \mathbb{Z}/2\mathbb{Z}).$$

Definition 6.2.21. Let $w : H_1(M; \mathbb{Z}) \to \mathbb{Z}/2\mathbb{Z}$ be the orientation character as in Lemma 6.2.18. Let $w_1(M) = e^{-1}(w) \in H^1(M; \mathbb{Z}/2\mathbb{Z})$. Then $w_1(M)$ is the first *Stiefel-Whitney* class of M. $\Diamond$

Corollary 6.2.22. *Let M be a manifold. Then M is orientable if and only if $w_1(M) = 0$.*

As usual, we do not just want to investigate objects but also maps between them. Here our objects are manifolds and the relevant maps are homeomorphisms.

Definition 6.2.23. Let M be a manifold and let $\{\overline{\varphi}_x\}$ be a compatible system of local G-orientations of M, thus giving a G-orientation of M, for $G = \mathbb{Z}/2\mathbb{Z}$ or $\mathbb{Z}$. Let $f : M \to N$ be a homeomorphism. Then the *induced G-orientation* on N is the G-orientation given by the compatible system of local G-orientations $\{\overline{\varphi}_y\}$ given as follows: For $y \in N$, let $x = f^{-1}(y)$. Then $\overline{\varphi}_y$ is the map making the diagram

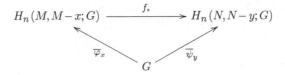

commute. $\Diamond$

Now we generalize from manifolds to manifolds with boundary. The first step could not be easier.

Definition 6.2.24. Let M be a manifold with boundary. M is orientable if $\operatorname{int}(M)$ is orientable. In that case, an orientation of M is an orientation of $\operatorname{int}(M)$. $\Diamond$

What is considerably more subtle is to see how an orientation of M gives an orientation of ∂M. We investigate that now.

We must see how to obtain local orientations on ∂M from local orientations on $\text{int}(M)$. The procedure is rather involved. We begin by working in $\mathbb{R}^n$, and we identify $\mathbb{R}^{n-1}$ with $\{(x_1,\ldots,x_{n-1},0)\} \subset \mathbb{R}^n_+$. We let $p = (0,\ldots,0) \in \mathbb{R}^n$ and $q = (0,\ldots,0,1) \in \mathbb{R}^n$. We let $C = \{x \in \mathbb{R}^{n-1} \mid |x| < 1\}$ and $\overline{B} = C \times [0,2) \subset \mathbb{R}^n_+$, and $B = C \times (0,2) \subseteq \overline{B}$.

Here is a schematic picture.

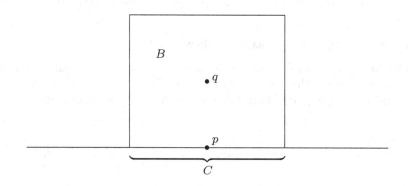

The key step is to construct an isomorphism

$$\tau : H_n(B, B - q) \longrightarrow H_{n-1}(C, C - p).$$

First we note that the boundary map in the exact sequence of the pair $(B, B - q)$ gives an isomorphism $H_n(B, B - q) \to H_{n-1}(B - q)$, and similarly the boundary map in the exact sequence of the pair $(C, C - p)$ gives an isomorphism $H_{n-1}(C, C - p) \to H_{n-2}(C - p)$. Now the inclusion $(B, B - q) \to (\overline{B}, \overline{B} - q)$ is a homotopy equivalence of pairs so induces an isomorphism on homology.

Now $\overline{B} - q$ is the union $\overline{B} - q = D \cup E$ where

$$D = \{(x_1,\ldots,x_n) \in \overline{B} \mid (x_1,\ldots,x_n) \neq (0,\ldots,0,t) \text{ for } 0 \leq t \leq 1\}$$

and

$$E = \{(x_1,\ldots,x_n) \in \overline{B} \mid 0 \leq x_n < 1\} = C \times [0,1).$$

Note that

$$D \cap E = (C - p) \times [0,1)$$

and the inclusion $C - p \to D \cap E$ is a homotopy equivalence, hence an isomorphism on homology.

We now consider the Mayer-Vietoris sequence for $\overline{B} - q = D \cup E$ and observe that we have an isomorphism

$$\partial : H_{n-1}(\overline{B} - q) \longrightarrow H_{n-2}(D \cap E).$$

Composing all these isomorphisms or their inverses as indicated below we obtain the isomorphism τ:

$$
\begin{array}{ccc}
H_n(B,B-q) \longrightarrow H_{n-1}(B-q) \longrightarrow H_{n-1}(\overline{B}-q) \\
\Big\downarrow \partial \\
H_{n-1}(C,C-p) \longleftarrow H_{n-2}(C-p) \longleftarrow H_{n-2}(D\cap E)
\end{array}
$$

Given this construction we make the following definition.

Definition 6.2.25. Let $\{\overline{\varphi}_y\}$ be an orientation of $\mathrm{int}\,(M)$, i.e., a compatible system of local orientations. The *induced orientation* of ∂M is the compatible system of local orientations $\{\overline{\varphi}_x\}$ obtained as follows: $\overline{\varphi}_x$ is the composition of isomorphisms and their inverses

$$
\mathbb{Z} \xrightarrow{\overline{\varphi}_y} H_n(M,M-y) \longleftarrow H_n(\varphi_\alpha(B),\varphi_\alpha(B)-y) \xleftarrow{(\varphi_\alpha)_*} H_n(B,B-q)
$$
$$
\Big\downarrow \tau
$$
$$
H_{n-1}(\partial M,\partial M-x) \longleftarrow H_{n-1}(\varphi_\alpha(C),\varphi_\alpha(C)-x) \xleftarrow{(\varphi_\alpha)_*} H_{n-1}(C,C-p)
$$

Here $\varphi_\alpha : \mathbb{R}^n_+ \to U_\alpha \subseteq M$ is a coordinate patch with $\varphi_\alpha(p) = x$ and $\varphi_\alpha(q) = y$. The maps labelled $(\varphi_\alpha)_*$ are both restrictions of φ_α to the respective domains, and the unlabelled maps are excision isomorphisms.   ◇

Theorem 6.2.26. *Let M be an oriented manifold with boundary. Then ∂M has a well-defined induced orientation given by the construction in Definition 6.2.25.*

In particular, ∂M is orientable.

Proof. This is simply a matter of checking that the local orientations $\{\overline{\varphi}_x\}$ are indeed compatible, and that they are independent of the choice of coordinate patches $(U_\alpha,\varphi_\alpha)$ used in the construction.   □

Remark 6.2.27. If M is not orientable, then ∂M may or may not be orientable (i.e., both possibilities may arise).   ◇

In practice, we often also want to consider (co)homology with coefficients in a field. In this regard we have the following result.

Lemma 6.2.28. *Let $G = \mathbb{F}$ be a field of characteristic 0 or odd characteristic. Then a manifold M is G-orientable if and only if it is orientable. If $G = \mathbb{F}$ is a field of characteristic 2, then every manifold M is G-orientable.*

Proof. We do the more interesting case of a field $\mathbb{F}$ of characteristic $\neq 2$. Consider the diagram in Definition 6.2.3.

If we let $V = \varphi_x(D)$ and replace G in that diagram by $H_n(M,M-V;G)$ and the two vertical maps by the isomorphisms induced by inclusions, we obtain a commutative diagram. This is true whether we use $\mathbb{F}$ coefficients or $\mathbb{Z}$ coefficients. But we also have the commutative diagram with the horizontal maps induced by the map $\mathbb{Z} \to \mathbb{F}$ of coefficients

$$
\begin{array}{ccc}
H_n(M, M - x; \mathbb{Z}) & \longrightarrow & H_n(M, M - x; \mathbb{F}) \\
\uparrow \cong & & \uparrow \cong \\
H_n(M, M - V; \mathbb{Z}) & \longrightarrow & H_n(M, M - V; \mathbb{F}) \\
\downarrow \cong & & \downarrow \cong \\
H_n(M, M - y; \mathbb{Z}) & \longrightarrow & H_n(M, M - y; \mathbb{F})
\end{array}
$$

Now if M is $\mathbb{Z}$-orientable it has a compatible collection of local $\mathbb{Z}$-orientations $\{\overline{\varphi}_x\}$ and then $\{\overline{\varphi}_x \otimes 1\}$ is a compatible collection of local $\mathbb{F}$-orientations.

On the other hand, suppose we have a compatible collection $\{\overline{\psi}_x\}$ of local $\mathbb{F}$-orientations. Fix a point $x \in M$. Then $\overline{\psi}_x(1) \in H_n(M, M - x; \mathbb{F})$ is a generator, i.e., a nonzero element. This element may not be in the image of $H_n(M, M - x; \mathbb{Z})$. But there is a nonzero element f of $\mathbb{F}$ (in fact, exactly two such) such that $f\overline{\psi}_x(1)$ is the image in $H_n(M, M - x; \mathbb{F})$ of a generator of $H_n(M, M - x; \mathbb{Z})$. By the commutativity of the above diagram that implies the same is true for $f\overline{\psi}_y(1)$ for every $y \in M$. Hence $\{f\overline{\psi}_x\}$ is a compatible system of local $\mathbb{Z}$-orientations of M, where x varies over M. $\qquad \square$

The proof of this lemma also shows how to obtain $\mathbb{F}$-orientations.

Definition 6.2.29. Let M be orientable and let $\mathbb{F}$ be an arbitrary field. An $\mathbb{F}$-orientation of M is a compatible system of local $\mathbb{F}$-orientations of the form $\{\overline{\varphi}_x \otimes 1\}$ where $\{\overline{\varphi}_x\}$ is a compatible system of local $\mathbb{Z}$-orientations of M.

Let M be arbitrary and let $\mathbb{F}$ be a field of characteristic 2. An $\mathbb{F}$-orientation of M is a compatible system of local $\mathbb{F}$-orientations of the form $\{\overline{\varphi}_x \otimes 1\}$ where $\{\overline{\varphi}_x\}$ is a compatible system of local $\mathbb{Z}/2\mathbb{Z}$-orientations of M. $\qquad \Diamond$

(In the characteristic 2 case compatibility is automatic.)

It is easy to check that the two parts of this definition agree when M is orientable and $\mathbb{F}$ has characteristic 2.

Orientability has very important homological implications, given by the following theorem. We state this theorem for manifolds with boundary, which includes the case of manifolds by taking $\partial M = \emptyset$. The hypothesis that M be connected is not essentially restrictive, as otherwise we could consider each component of M separately.

Theorem 6.2.30. *Let M be a compact connected n-manifold with boundary. Let $G = \mathbb{Z}/2\mathbb{Z}$ or $\mathbb{Z}$.*

If M is G-oriented, suppose that $\{\overline{\varphi}_x\}$ is a compatible system of local G-orientations giving the G-orientation of M. In this case, there is a unique *homology class* $[M,\partial M] \in H_n(M,\partial M;G)$ *with* $i_*([M,\partial M]) = \varphi_x(1) \in H_n(M,M-x;G)$ *for every* $x \in M$, *where* $i : (M,\partial M) \to (M,M-x)$ *is the inclusion of pairs. Furthermore,* $[M,\partial M]$ *is a generator of* $H_n(M,\partial M;G)$.

If M is not G-orientable, then $H_n(M,\partial M;G) = 0$.

Since this theorem is so important, we will explicitly state one of its immediate consequences.

Corollary 6.2.31. *Let M be a compact connected n-manifold with boundary.*

(1) *For any such M,* $H_n(M,\partial M;\mathbb{Z}/2\mathbb{Z}) \cong \mathbb{Z}/2\mathbb{Z}$ *and* $H^n(M,\partial M;\mathbb{Z}/2\mathbb{Z}) \cong \mathbb{Z}/2\mathbb{Z}$.
(2) *If M is orientable, then* $H_n(M,\partial M;\mathbb{Z}) \cong \mathbb{Z}$ *and* $H^n(M,\partial M;\mathbb{Z}) \cong \mathbb{Z}$. *If M is not orientable, then* $H_n(M,\partial M;\mathbb{Z}) = 0$ *and* $H^n(M,\partial M;\mathbb{Z}) = 0$.

Proof. The statements on homology are a direct consequence of Theorems 6.2.6 and 6.2.30.

The statements for cohomology then follow from the universal coefficient theorem and Theorem 6.1.12. □

Example 6.2.32. We computed the homology of $\mathbb{R}P^n$ in Theorem 4.3.4. Combining that result with Corollary 6.2.31, we see that $\mathbb{R}P^n$ is orientable for n odd and nonorientable for n even. ◇

Definition 6.2.33. Let M be a compact connected G-oriented n-manifold, $G = \mathbb{Z}/2\mathbb{Z}$ or $\mathbb{Z}$. The homology class $[M,\partial M] \in H_n(M,\partial M;G)$ as in Theorem 6.2.30 is called the *fundamental homology class* (or simply *fundamental class*) of $(M,\partial M)$. Its dual $\{M,\partial M\}$ in $H^n(M,\partial M;G)$, i.e., the cohomology class with $e(\{M,\partial M\},[M,\partial M]) = 1$, is the *fundamental cohomology class* of $(M,\partial M)$.

If M is oriented and G is any coefficient group, the image of $[M,\partial M]$ in $H_n(M,\partial M;G)$ under the coefficient map $\mathbb{Z} \to G$ is also called a fundamental homology class, and similarly the image of $\{M,\partial M\}$ on $H^n(M,\partial M;G)$ under the same coefficient map is also called a fundamental cohomology class. ◇

Remark 6.2.34. We need to be careful in Definition 6.2.33 when we referred to the dual of a homology class. In general, if V is a free abelian group (or a vector space) if does not make sense to speak of the dual of an element v of V. But it does make sense here. Suppose that M is connected. Then $H_n(M;G)$ is free of rank 1 (in case M is orientable and $G = \mathbb{Z}$) or is a 1-dimensional vector space over $\mathbb{Z}/2\mathbb{Z}$ (for M arbitrary and $G = \mathbb{Z}/2\mathbb{Z}$) and we have the pairing $e : H^n(M;G) \otimes H_n(M;G) \to G$. In this situation, given a generator v of $H_n(M;G)$, there is a unique element (also a generator) v^* in $H^n(M;G)$ with $e(v^*,v) = 1$, and v^* is what we mean by the dual of v.

In general, if M is the disjoint union of k components, $M = M_1 \cup \cdots \cup M_n$, then $[M]$ is the sum $[M_1] + \cdots + [M_k] \in H_n(M;G)$, and we take $\{M\}$ to be the sum $\{M_1\} + \cdots + \{M_k\} \in H^n(M;G)$. ◇

Recall we showed in Theorem 6.2.26 that if M is an orientable n-manifold with boundary ∂M, then ∂M is also orientable, and furthermore we showed how to obtain an orientation for ∂M from that on M. That fits in with the homological description as follows.

Corollary 6.2.35. *Let M be a G-oriented n-manifold with boundary, $G = \mathbb{Z}/2\mathbb{Z}$ or $\mathbb{Z}$, given by a compatible system of local orientations on M, and let $[M, \partial M] \in H_n(M, \partial M; G)$ be the fundamental class of M with this orientation. Let ∂M have the induced orientation as in Definition 6.2.25, and let $[\partial M] \in H_{n-1}(\partial M; G)$ be the fundamental class of ∂M with this orientation. Then*

$$[\partial M] = \partial([M, \partial M])$$

where $\partial : H_n(M, \partial M; G) \to H_{n-1}(\partial M; G)$ is the boundary map in the exact sequence of the pair $(M, \partial M)$.

We record the following observation for future use.

Corollary 6.2.36. *Let M be a G-oriented n-manifold with boundary with fundamental class $[M, \partial M]$, and let ∂M have the induced G-orientation with fundamental class $[\partial M]$. If $i : \partial M \to M$ is the inclusion, then $i_*([\partial M]) = 0 \in H_{n-1}(M; G)$.*

Proof. We have the exact sequence of the pair $(M, \partial M)$:

$$H_n(M, \partial M; G) \xrightarrow{\ \partial\ } H_{n-1}(\partial M; G) \xrightarrow{\ i_*\ } H_{n-1}(M; G).$$

But $[\partial M] = \partial([M, \partial M])$, so by exactness $i_*([\partial M]) = 0$. $\qquad\square$

6.3 Examples of Orientability and Nonorientability

In this section we look at a family of manifolds, and by directly working with local orientations and coordinate patches determine when they are orientable.

Remark 6.3.1. We have properly defined an orientation of M as a compatible system of local orientations $\{\overline{\varphi}_x\}$, where $\overline{\varphi}_x : \mathbb{Z} \to H_n(M, M - x; \mathbb{Z})$ is an isomorphism. This gives a choice of generator $\overline{\varphi}_x(1) \in H_n(M, M - x; \mathbb{Z})$. It is common to think of an orientation of M as a consistent choice of generators of $H_n(M, M - x; \mathbb{Z})$ for every $x \in M$, where a consistent choice of generators means that $\{\overline{\varphi}_x\}$ are all compatible. $\diamond$

We now give several examples of orientable and nonorientable manifolds (and we will be using the approach of this remark in our investigation).

Example 6.3.2. (1) We shall show that S^1 is orientable. We take a rather strange looking description and parameterization of S^1, but we do so to use this as a "warm-up" for the next part of this example, where the corresponding parameterization will simplify matters. We let

$$I_\alpha = (-2, 6) \quad \text{and} \quad I_\beta = (-6, 2)$$

with I_α parameterized by s and I_β parameterized by t, and we let M^1 be the identification space

$$I_\alpha \cup I_\beta / \sim$$

where the identification $\sim$ is given by

$$s \in I_\alpha \sim s \in I_\beta \qquad \text{if } -2 < s < 2$$

$$s \in I_\alpha \sim s - 8 \in I_\beta \quad \text{if } 2 < s < 6.$$

Then M^1 is homeomorphic to S^1. A homeomorphism is given by $s \mapsto \exp(\pi i s/4)$. (Note this map is well-defined on M^1.)

We let $\varphi_\alpha : I_\alpha \to M^1$ and $\varphi_\beta : I_\beta \to M^1$ be the inclusions and let $U_\alpha = \varphi_\alpha(I_\alpha)$, $U_\beta = \varphi_\beta(I_\beta)$. Then $\{(U_\alpha, \varphi_\alpha), (U_\beta, \varphi_\beta)\}$ is an atlas on M^1.

Let $f(r) = \varphi_\alpha(r)$, $0 \le r \le 4$. Then $f(r)$ is a path from $x = f(0)$ to $y = f(1)$ lying entirely in U_α. Let $\overline{g}(r) = \varphi_\beta(r-8)$, $4 \le r \le 8$. Then $\overline{g}(r)$ is a path from $\overline{g}(4)$ to $\overline{g}(8)$ lying entirely in U_β. But $\overline{g}(4) = f(1)$ and $\overline{g}(8) = f(0)$, so

$$h(r) = \begin{cases} f(r) & 0 \le r \le 4 \\ \overline{g}(r) & 4 \le r \le 8 \end{cases}$$

is a path in M^1 from x to x. It is easy to check that under the above homeomorphism, this becomes a loop in S^1 from 1 to 1 winding once around S^1 counterclockwise. In particular this loop represents a generator of $\pi_1(M^1, x)$, so to show that M^1 is orientable we need only show that $h_x(\overline{\varphi}_x) = \overline{\varphi}_x$, where $\overline{\varphi}_x$ is a local orientation of M^1 at x.

We need to establish a bit of notation. Recall that for a space consisting of the point z we have the generator 1_z of $H_0(z)$ as in Definition 5.1.21. That notation would be too confusing in the current context, so instead we denote this homology class by $\langle z \rangle$.

Recall that if I is an open interval and i is any point of I, we have the isomorphism $\partial : H_1(I, I-i) \to \overline{H}_0(I-i)$ of the exact sequence of the pair $(I, I-i)$.

We have the generator $\partial^{-1}(\langle 1 \rangle - \langle -1 \rangle)$ of $H_1(I_\alpha, I_\alpha - \{0\})$ and we give M^1 the local orientation at x determined by

$$\overline{\varphi}_x(1) = (\varphi_\alpha)_*(\partial^{-1}(\langle 1 \rangle - \langle -1 \rangle)) \in H_1(M, M-x).$$

Now the points x and y both lie in the same coordinate path U_γ for both $\gamma = \alpha$ and $\gamma = \beta$. Let $p = \varphi_\gamma^{-1}(x)$ and $q = \varphi_\gamma^{-1}(y)$. Then we have the diagram from Definition 6.2.25 (where we omit the intermediate groups)

$$H_1(M^1, M^1 - x) \xleftarrow{\ (\varphi_\gamma)_* \ } H_1(I_\gamma, I_\gamma - p)$$

$$\Big\uparrow$$

$$H_1(I_\gamma, I_\gamma - D)$$

$$\Big\downarrow$$

$$H_1(M^1, M^1 - y) \xleftarrow{\ (\varphi_\gamma)_* \ } H_1(I_\gamma, I_\gamma - q)$$

Since the path f lies entirely in U_α, we use this diagram with $\gamma = \alpha$ to compute $f_y(\overline{\varphi}_x)$.

$$
\begin{aligned}
((\varphi_\alpha)_*)^{-1}(\overline{\varphi}_x(1)) &= \partial^{-1}(\langle 1 \rangle - \langle -1 \rangle) \in H_1(I_\alpha, I_\alpha - \{0\}) \\
&= \partial^{-1}(\langle 5 \rangle - \langle -1 \rangle) \in H_1(I_\alpha, I_\alpha - \{0\}) \\
&= \partial^{-1}(\langle 5 \rangle - \langle -1 \rangle) \in H_1(I_\alpha, I_\alpha - [0,4]) \\
&= \partial^{-1}(\langle 5 \rangle - \langle -1 \rangle) \in H_1(I_\alpha, I_\alpha - \{4\}) \\
&= \partial^{-1}(\langle 5 \rangle - \langle 3 \rangle) \in H_1(I_\alpha, I_\alpha - \{4\})
\end{aligned}
$$

giving the local orientation $\overline{\varphi}_y$ at y specified by

$$\overline{\varphi}_y(1) = (\varphi_\alpha)_*(\partial^{-1}(\langle 5 \rangle - \langle 3 \rangle)) \in H_1(M^1, M^1 - y).$$

Similarly, the path $\overline{g}$ lies entirely in U_β, so we use the diagram with $\gamma = \beta$ to compute $\overline{g}_x(\overline{\varphi}_y)$. The key to this computation is starting it off correctly.

$$
\begin{aligned}
((\varphi_\beta)_*)^{-1}(\overline{\varphi}_y(1)) &= ((\varphi_\beta)_*)^{-1}(\varphi_\alpha)_*(\partial^{-1}(\langle 5 \rangle - \langle 3 \rangle)) \\
&= (\varphi_\beta^{-1}\varphi_\alpha)_*(\partial^{-1}(\langle 5 \rangle - \langle 3 \rangle)).
\end{aligned}
$$

Referring to our original identifications, we see that

$$5 \in I_\alpha \sim -3 \in I_\beta, \quad 3 \in I_\alpha \sim -5 \in I_\beta$$

and so

$$
\begin{aligned}
((\varphi_\beta)_*)^{-1}(\overline{\varphi}_y(1)) &= \partial^{-1}(\langle -3 \rangle - \langle -5 \rangle) \in H_1(I_\beta, I_\beta - \{-4\}) \\
&= \partial^{-1}(\langle 1 \rangle - \langle -5 \rangle) \in H_1(I_\beta, I_\beta - \{-4\}) \\
&= \partial^{-1}(\langle 1 \rangle - \langle -5 \rangle) \in H_1(I_\beta, I_\beta - [-4,0]) \\
&= \partial^{-1}(\langle 1 \rangle - \langle -5 \rangle) \in H_1(I_\beta, I_\beta - \{0\}) \\
&= \partial^{-1}(\langle 1 \rangle - \langle -1 \rangle) \in H_1(I_\beta, I_\beta - \{0\})
\end{aligned}
$$

giving the local orientation $\overline{\varphi}'_x$ at x specified by

$$\overline{\varphi}'_x(1) = (\varphi_\beta)_*(\partial^{-1}(\langle 1 \rangle - \langle -1 \rangle)) \in H_1(M^1, M^1 - x).$$

Now M^1 is orientable if and only if this new local orientation $\overline{\varphi}'_x$ at x is the same as the original local orientation. But, again referring to our original identifications, we see that $1 \in I_\alpha \sim 1 \in I_\beta$ and $-1 \in I_\alpha \sim -1 \in I_\beta$ so this is indeed the case.

(2) We now do a more elaborate, and much more interesting, computation along the same lines. Let m be a non negative integer and let $n = m + 1$. Let

$$V_\alpha = (-2,6) \times (-4,4)^m \quad \text{and} \quad V_\beta = (-6,2) \times (-4,4)^m$$

(where again these numbers are just chosen for convenience) and let M^n be the identification space

$$M^n = V_\alpha \cup V_\beta / \sim$$

where the identification is as follows:

$$(s,t) \in V_\alpha \sim (s,t) \in V_\beta \quad \text{if } -2 < s < 2$$
$$(s,t) \in V_\alpha \sim (s-8,-t) \in V_\beta \quad \text{if } 2 < s < 6.$$

Here $t = (t_1, \ldots, t_m)$ is an m-tuple of real numbers and $-t = (-t_1, \ldots, -t_m)$. In case $m = 0$, we just recover S^1, as in part (1). In case $m = 1$, M^n is a Möbius strip (or, to be precise, an "open" Möbius strip, as its boundary is missing). In general, M^n is an n-manifold.

We shall show that M^n is orientable if m is even (i.e., if n is odd) and that M^n is nonorientable if m is odd (i.e., if n is even). In particular, the Möbius strip is nonorientable.

We let $\varphi_\alpha : V_\alpha \to M^n$ and $\varphi_\beta : V_\beta \to M^n$ be the inclusions, and let $U_\alpha = \varphi_\alpha(V_\alpha)$ and $U_\beta = \varphi_\beta(V_\beta)$. Thus $\{(U_\alpha, \varphi_\alpha), (U_\beta, \varphi_\beta)\}$ is an atlas for M^n.

We begin by observing that M^n has $\{(t,0)\}$ as a strong deformation retract, and $\{(t,0)\}$ is just S^1 (up to homeomorphism). Thus, as in part (1), to check whether M^n is orientable we just have to consider what happens on a single loop that generates the fundamental group of M^n.

We let $x = \varphi_\alpha(0,0) = \varphi_\beta(0,0)$ and $y = \varphi_\alpha(4,0) = \varphi_\beta(-4,0)$. We let $f(r) = \varphi_\alpha(r,0)$, $0 \le r \le 4$ and $\overline{g}(r) = \varphi_\beta(r-8,0)$, $4 \le r \le 8$ so that

$$h(r) = \begin{cases} f(r) & 0 \le r \le 4 \\ \overline{g}(r) & 4 \le r \le 8 \end{cases}$$

is a loop in M^n that represents a generator of $\pi_1(M^n, x)$.

Again we observe that if $V = V_\alpha$ or V_β and v is any point of V, we have the isomorphism $\partial : H_n(V, V - v) \to H_{n-1}(V - v)$ of the exact sequence of the pair $(V, V - v)$.

Let $i : S^{n-1} \to V_\alpha - (0,0)$ by $i(s, t_1, \ldots, t_m) = (s, t_1, \ldots, t_m)$ and choose a generator g_0 of $H_{n-1}(S^{n-1})$ so that $i_*(g_0) = g_1$ is a generator of $H_{n-1}(V_\alpha - (0,0))$.

Observe that i is homotopic to $j : S^{n-1} \to V_\alpha - (0,0)$ by $j(s, t_1, \ldots, t_m) = (3(s+2), t_1, \ldots, t_m)$, so that $g_2 = j_*(g_0) = g_1 \in H_{n-1}(V_\alpha - (0,0))$, and that $j(S^{n-1}) \subset V_\alpha - D$ where D is the disk of radius 2 around $(2,0)$, so that $(0,0)$ and $(4,0)$ are both contained in D. The inclusion $V - D \to V - (0,0)$ induces an isomorphism on homology and we identify $H_{n-1}(V - (0,0))$ and $H_{n-1}(V - D)$ by this isomorphism, and similarly for $H_{n-1}(V - (4,0))$ and $H_{n-1}(V - D)$. In turn j is homotopic to $k : S^{n-1} \to V_\alpha - (4,0)$ by $k(s, t_1, \ldots, t_m) = (s+4, t_1, \ldots, t_m)$ so that $g_3 = k_*(g_0) = g_2 \in H_{n-1}(V - (4,0))$.

With these observations out of the way, we get down to work. The argument here parallels that in part (1). We begin by giving M^n the local orientation at x determined by

$$\overline{\varphi}_x(1) = (\varphi_\alpha)_*(\partial^{-1}(g_1)) \in H_n(M^n, M^n - x).$$

Since the path f lies entirely in U_α, we may use the analogous method to compute $f_y(\overline{\varphi}_x)$, translating everything to V_α. We then get

$$((\varphi_\alpha)_*)^{-1}(\overline{\varphi}_x(1)) = \partial^{-1}(g_1) \in H_n(V_\alpha, V_\alpha - (0,0))$$
$$= \partial^{-1}(g_2) \in H_n(V_\alpha, V_\alpha - (0,0))$$
$$= \partial^{-1}(g_2) \in H_n(V_\alpha, V_\alpha - D)$$
$$= \partial^{-1}(g_2) \in H_n(V_\alpha, V_\alpha - (4,0))$$
$$= \partial^{-1}(g_3) \in H_n(V_\alpha, V_\alpha - (4,0))$$

giving the local orientation $\overline{\varphi}_y$ at y specified by

$$\overline{\varphi}_y(1) = (\varphi_\alpha)_*(\partial^{-1}(g_3)) \in H_n(M^n, M^n - y).$$

Now we translate everything to V_β to compute $\overline{g}_x(\overline{\varphi}_y)$. Again it is crucial to start correctly. We have $g_3 = k_*(g_0)$, so

$$((\varphi_\beta)_*)^{-1}(\overline{\varphi}_y(1)) = (\varphi_\beta)_*^{-1}(\varphi_\alpha)_*(\partial^{-1}(k_*(g_0)))$$
$$= \partial^{-1}(\varphi_\beta^{-1}\varphi_\alpha k)_*(g_0).$$

Now on the one hand, we have $k : S^{n-1} \to V_\alpha$ by

$$k(s,t) = (s+4,t)$$

and on the other hand, we may consider $l : S^{n-1} \to V_\beta$ by

$$l(s,t) = (s-4,t).$$

But, referring to our original identifications, we see that $(s+4,t) \sim (s-4,-t)$, i.e., that $\varphi_\alpha k = \varphi_\beta l$. Hence we see that

$$\varphi_\beta^{-1} \varphi_\alpha k = l.$$

Now we follow the same argument to conclude that we obtain a new local orientation of M^n at x given by $\overline{\varphi}'_x(1) = \overline{g}_x(\overline{\varphi}_y(1)) = (\varphi_\beta)_*(\partial^{-1}(g'_1))$ where $g'_1 = m_*(g_0)$ with $m(s,t) = (s,-t)$. But in a neighborhood of $(0,0)$, the identification is $(s,0) \sim (s,0)$, i.e., $\varphi_\beta = \varphi_\alpha$ there. Thus $\overline{\varphi}'_x(1) = \pm\overline{\varphi}_x(1)$ according as $g'_1 = m_*(g_0) = \pm g_1 = \pm i_*(g_0)$.

To determine the sign, we observe that m is the composition $m = \overline{i}\overline{a}$ where $\overline{a} : S^{n-1} \to S^{n-1}$ by $\overline{a}(s,t) = (s,-t)$. We recognize $\overline{a}$ as the suspension of the antipodal map $a : S^{m-1} \to S^{m-1}$, except that we are using the first coordinate as the suspension coordinate rather than the last one, so the degree of $\overline{a}$ is equal to the degree of a, which is $(-1)^m$ by Lemma 4.1.10.

(3) There is a homeomorphism $h : M^n \to N^n \subset \mathbb{R}P^n$ of M^n onto an open subset of $\mathbb{R}P^n$ for each n, which includes an isomorphism $h_* : H_1(M^n; \mathbb{Z}/2\mathbb{Z}) \to H_1(\mathbb{R}P^n; \mathbb{Z}/2\mathbb{Z})$ and so we may compute the orientation character of $\mathbb{R}P^n$ by considering M^n. Thus we conclude from part (2) that $\mathbb{R}P^n$ is orientable for n odd and nonorientable for n even. $\Diamond$

Remark 6.3.3. Our conclusion in Example 6.3.2(3) of course agrees with our conclusion in Example 6.2.32. $\Diamond$

6.4 Poincaré and Lefschetz Duality and Applications

In this section we give the basic duality theorems: Poincaré duality for compact G-oriented manifolds and Lefschetz duality for compact G-oriented manifolds with boundary. We then do some examples and give some important applications of these important theorems. Throughout, we let $G = \mathbb{Z}$ or a field $\mathbb{F}$.

First we deal with the case of compact manifolds. Let M be a compact G-oriented n-manifold. Recall we defined the fundamental homology class $[M] \in H_n(M; G)$ in Definition 6.2.33.

Theorem 6.4.1 (Poincaré duality). *Let M be a compact G-oriented n-manifold, and let $[M] \in H_n(M; G)$ be its fundamental homology class. Then the map*

$$\cap [M] : H^j(M; G) \longrightarrow H_{n-j}(M; G)$$

is an isomorphism for every $j = 0, \dots, n$.

Now we deal with compact manifolds with boundary. Again for a compact G-oriented n-manifold M with boundary we have the fundamental homology class $[M, \partial M] \in H_n(M, \partial M; G)$, and in this situation $\partial([M, \partial M]) = [\partial M]$ is the fundamental homology class of ∂M with the induced G-orientation.

Theorem 6.4.2 (Lefschetz duality). *Let M be a compact G-oriented n-manifold with boundary. In the diagram*

$$\cdots \longrightarrow H^{j-1}(M;G) \longrightarrow H^{j-1}(\partial M;G) \longrightarrow H^{j}(M,\partial M;G) \longrightarrow H^{j}(M;G) \longrightarrow \cdots$$
$$\Big\downarrow{\scriptstyle \cap[M,\partial M]} \qquad \Big\downarrow{\scriptstyle \cap[\partial M]} \qquad \Big\downarrow{\scriptstyle \cap[M,\partial M]} \qquad \Big\downarrow{\scriptstyle \cap[M,\partial M]}$$
$$\cdots \longrightarrow H_{n-j+1}(M,\partial M;G) \longrightarrow H_{n-j}(\partial M;G) \longrightarrow H_{n-j}(M;G) \longrightarrow H_{n-j}(M,\partial M;G) \longrightarrow \cdots$$

the left-hand square commutes up to a factor of $(-1)^{j-1}$, the middle and right-hand squares commute, and the vertical maps are all isomorphisms.

In particular, the maps

$$\cap[M,\partial M] : H^{j}(M,\partial M;G) \longrightarrow H_{n-j}(M;G)$$

and

$$\cap[M,\partial M] : H^{j}(M;G) \longrightarrow H_{n-j}(M,\partial M;G)$$

are isomorphisms for every $j = 0,\ldots,n$.

We now give several examples of how to use Poincaré duality to compute the structure of cohomology rings. In these examples, we will be using the properties of cup and cap products given in Theorem 5.6.13, and the notation in that theorem.

Example 6.4.3. (1) Let $G = \mathbb{Z}$. Choose an orientation of S^p, and let S^p have fundamental homology class $[S^p]$ and fundamental cohomology class $\{S^p\}$. Also choose an orientation of S^q and let S^q have fundamental homology class $[S^q]$ and fundamental cohomology class $\{S^q\}$.

Let $M = S^p \times S^q$ with $p, q \geq 1$, so that M is connected. For simplicity, we consider the case $p \neq q$. (The argument for $p = q$ is similar but a bit more complicated.) Then $H_p(M)$ is generated by $\tilde{a} = (i_1)_*([S^p])$ and $H_q(M)$ is generated by $\tilde{b} = (i_2) * ([S^q])$ where $i_1 : S^p \to M$ and $i_2 : S^q \to M$ are the inclusions $S^p \to S^p \times \{*\} \subset M$ and $S^q \to \{*\} \times S^q \subset M$. Also, $H^p(M)$ is generated by $\tilde{\alpha} = \pi_1^*(\{S^p\})$ and $H^q(M)$ is generated by $\tilde{\beta} = \pi_2^*(\{S^q\})$ where $\pi_1 : M \to S^p$ and $\pi_2 : M \to S^q$ are projections onto the first and second factors respectively.

By Poincaré duality, $\cap[M] : H^q(M) \to H_p(M)$ is an isomorphism. Hence $\tilde{\beta} \cap [M] = \pm \tilde{a}$, and we choose the orientation on M so that the sign is positive. Then

$$1 = e(\tilde{\alpha}, \tilde{a}) = e(\tilde{\alpha}, \tilde{\beta} \cap [M]) = e(\tilde{\alpha} \cup \tilde{\beta}, [M]) = e(\{M\}, [M])$$

and hence $\tilde{\gamma} = \tilde{\alpha} \cup \tilde{\beta} = \{M\}$ is a generator of $H^{p+q}(M)$. We thus recover the computation of Example 5.7.1.

(2) We take $G = \mathbb{Z}/2\mathbb{Z}$. Let $\alpha \in H^1(\mathbb{R}P^n)$ be a generator. We claim that $\alpha^k \in H^k(\mathbb{R}P^n)$ is a generator for each k. We prove this by induction on n. The case $n = 1$ is trivial. Assume this is true for n and consider $\mathbb{R}P^{n+1}$. The inclusion $\mathbb{R}P^n \to \mathbb{R}P^{n+1}$ induces isomorphism on (co)homology in dimensions at most n. By Poincaré duality,

$$\cap[\mathbb{R}P^{n+1}] : H^1(\mathbb{R}P^{n+1}) \longrightarrow H_n(\mathbb{R}P^{n+1})$$

is an isomorphism, and so

$$\alpha \cap [\mathbb{R}P^{n+1}] = [\mathbb{R}P^n] \in H_n(\mathbb{R}P^{n+1}).$$

By the induction hypothesis, $\alpha^n = \{\mathbb{R}P^n\} \in H^n(\mathbb{R}P^{n+1})$. But then

$$1 = e(\{\mathbb{R}P^n\}, [\mathbb{R}P^n]) = e(\alpha^n, \alpha \cap [\mathbb{R}P^{n+1}])$$
$$= e(\alpha^{n+1}, [\mathbb{R}P^{n+1}]) = e(\{\mathbb{R}P^{n+1}\}, [\mathbb{R}P^{n+1}])$$

so $\alpha^{n+1} = \{\mathbb{R}P^{n+1}\}$. Thus α^{n+1} is a generator of $H^{n+1}(\mathbb{R}P^{n+1})$, and so α^k is a generator of $H^k(\mathbb{R}P^{n+1})$ for each $0 \le k \le n+1$ (as if for some j, α^j were not a generator of $H^j(\mathbb{R}P^{n+1})$, $\alpha^{n+1} = \alpha^j \cup \alpha^{n+1-j}$ could not be a generator of $H^{n+1}(\mathbb{R}P^{n+1})$). Thus we recover the calculation of Lemma 5.7.4(1).

(3) We take $G = \mathbb{Z}$. By a completely analogous argument, we recover the calculation of Lemma 5.7.4(2) for $H^*(\mathbb{C}P^n)$.   ◊

Example 6.4.4. Example 6.4.3(3) allows us to define a family of orientations on complex projective spaces.

We refer to the orientations obtained in this way as the standard orientations, and the ones with the opposite sign for $[\mathbb{C}P^n]$ as the nonstandard orientations.

We begin with $n = 1$. We have the standard generator $\sigma_1 \in H_1(S^1)$ of Remark 4.1.10.

To define an orientation of $\mathbb{C}P^1$ it suffices to give a local orientation $\overline{\varphi}_{z_0}$ at a single point z_0, and we choose z_0 to be the point with homogeneous coordinates $[0,1]$. We specify $\overline{\varphi}_{z_0}$ by letting $\overline{\varphi}_{z_0}(1)$ be the image of σ_1 under the sequence of isomorphisms

$$H_1(S^1) \longrightarrow H_1(\mathbb{C} - \{0\}) \longrightarrow H_2(\mathbb{C}, \mathbb{C} - \{0\}) \longrightarrow H_2(\mathbb{C}P^1, \mathbb{C}P^1 - \{[0,1]\}).$$

Here the first isomorphism is induced by inclusion, the second is the inverse of the boundary map in the exact sequence of the pair $(\mathbb{C}, \mathbb{C} - \{0\})$, and the third is induced by the map $z \mapsto [z, 1]$.

Given this orientation we have a fundamental class $[\mathbb{C}P^1] \in H_2(\mathbb{C}P^1)$, and we let $\alpha = \{\mathbb{C}P^1\}$ be the fundamental cohomology class. Then for $n > 1$, we choose the orientation which has $\{\mathbb{C}P^n\} = \alpha^n$ as fundamental cohomology class, i.e., the orientation with fundamental class $[\mathbb{C}P^1]$ specified by $e(\alpha^n, [\mathbb{C}P^n]) = 1$.

We then write $\mathbb{C}P^n$ for $\mathbb{C}P^n$ the oriented manifold with the standard orientation and $\overline{\mathbb{C}P^n}$ for $\mathbb{C}P^n$ the oriented manifold with the nonstandard orientation.

(The standard orientations may be obtained directly by specifying local orientations in a completely analogous manner, beginning with the standard generator $\sigma_{2n-1}(S^{2n-1})$, but for our purposes the homological description is much more to the point.)   ◊

Theorem 6.4.5. *Let M be a compact n-dimensional manifold with n odd. Then the Euler characteristic $\chi(M) = 0$.*

Proof. We may use any coefficients to compute the Euler characteristic, so we choose $\mathbb{Z}/2\mathbb{Z}$. This means that M is orientable with these coefficients. Also, they form a field, so for any j, $H_j(M; \mathbb{Z}/2\mathbb{Z})$ and $H^j(M; \mathbb{Z}/2\mathbb{Z})$ are dual vector spaces and hence have the same dimension. Let $n = 2m + 1$. We compute

$$\chi(M) = \sum_{j=0}^{n} (-1)^j \dim H_j(M; \mathbb{Z}/2\mathbb{Z})$$

$$= \sum_{j=0}^{m} (-1)^j \dim H_j(M; \mathbb{Z}/2\mathbb{Z}) + \sum_{k=m+1}^{2m+1} (-1)^k \dim H_k(M; \mathbb{Z}/2\mathbb{Z}).$$

But by Poincaré duality

$$\sum_{k=m+1}^{2m+1} (-1)^k \dim H_k(M; \mathbb{Z}/2\mathbb{Z}) = \sum_{k=m+1}^{2m+1} (-1)^k \dim H^{2m+1-k}(M; \mathbb{Z}/2\mathbb{Z})$$

$$= \sum_{k=m+1}^{2m+1} (-1)^k \dim H_{2m+1-k}(M; \mathbb{Z}/2\mathbb{Z})$$

$$= \sum_{j=0}^{m} (-1)^{2m+1-j} \dim H_j(M; \mathbb{Z}/2\mathbb{Z})$$

where $j = 2m + 1 - k$ (and so $k = 2m + 1 - j$).
Thus

$$\chi(M) = \sum_{j=0}^{m} ((-1)^j + (-1)^{2m+1-j}) \dim H_j(M; \mathbb{Z}/2\mathbb{Z}).$$

But $(-1)^j$ and $(-1)^{2m+1-j}$ always have opposite signs, so this sum is identically zero.   □

Now we have an interesting application of Lefschetz duality.

Theorem 6.4.6. *Let M be a compact n-manifold with odd Euler characteristic. Then M is not the boundary of a compact $(n+1)$-manifold.*

Proof. Suppose that M is the boundary of the compact $(n+1)$-manifold X. Consider the exact sequence of the pair (X,M):

$$0 \longrightarrow H_{n+1}(X;\mathbb{Z}/2\mathbb{Z}) \longrightarrow H_{n+1}(X,M;\mathbb{Z}/2\mathbb{Z}) \longrightarrow H_n(M;\mathbb{Z}/2\mathbb{Z}) \longrightarrow$$
$$\cdots \longrightarrow H_0(M;\mathbb{Z}/2\mathbb{Z}) \longrightarrow H_0(X;\mathbb{Z}/2\mathbb{Z}) \longrightarrow H_0(X,M;\mathbb{Z}/2\mathbb{Z}) \longrightarrow 0.$$

Then the alternating sum of the dimensions of the homology groups in this sequence is zero, and hence the sum of the dimensions is even. Thus

$$\sum_{k=0}^{n} \dim H_k(M;\mathbb{Z}/2\mathbb{Z}) + \sum_{k=0}^{n+1} \dim H_k(X;\mathbb{Z}/2\mathbb{Z})$$
$$+ \sum_{k=0}^{n+1} \dim H_k(X,M;\mathbb{Z}/2\mathbb{Z}) \equiv 0 \ (\mathrm{mod}\ 2).$$

But, on the other hand,

$$\chi(M) = \sum_{k=0}^{n} (-1)^k \dim H_k(M;\mathbb{Z}/2\mathbb{Z}) \equiv \sum_{k=0}^{n} \dim H_k(M;\mathbb{Z}/2\mathbb{Z}) \ (\mathrm{mod}\ 2),$$

and on the other hand, by Lefschetz duality,

$$\sum_{k=0}^{n+1} \dim H_k(X;\mathbb{Z}/2\mathbb{Z}) = \sum_{k=0}^{n+1} \dim H_k(X,M;\mathbb{Z}/2\mathbb{Z})$$

since for every value of k,

$$\dim H_k(X;\mathbb{Z}/2\mathbb{Z}) = \dim H^{n+1-k}(X,M;\mathbb{Z}/2\mathbb{Z}) = \dim H_{n+1-k}(X,M;\mathbb{Z}/2\mathbb{Z}).$$

Thus $\chi(M)$ is even, a contradiction. $\square$

Now we turn our attention to orientable manifolds. We will be considering an important invariant, the intersection form. In order to most conveniently do that we introduce some (nonstandard) notation.

Definition 6.4.7. Let M be a compact connected manifold of dimension $2n$. Then

$$K^n(M;\mathbb{Z}) = H^n(M;\mathbb{Z})/H^n(M;\mathbb{Z})_{\mathrm{tor}},$$

i.e., the quotient of $H^n(M;\mathbb{Z})$ by its torsion subgroup, and

$$K^n(M;\mathbb{F}) = H^n(M;\mathbb{F})$$

if $\mathbb{F}$ is a field. $\Diamond$

Theorem 6.4.8. *Let M be a G-oriented compact connected manifold of even dimension $2n$, with fundamental class $[M] \in H_{2n}(M;G)$, where $G = \mathbb{Z}$ or a field $\mathbb{F}$. Then*

$$\langle\,,\,\rangle : K^n(M;G) \otimes K^n(M;G) \longrightarrow G$$

given by

$$\langle u,v \rangle = e(u \cup v, [M])$$

is a nonsingular bilinear form. It is symmetric if n is even, i.e., if $\dim(M) \equiv 0 \pmod 4$, and is skew-symmetric if n is odd, i.e., if $\dim(M) \equiv 2 \pmod 4$.

Proof. First observe that this form is symmetric for n even and skew-symmetric for n odd as we have $u \cup v = (-1)^{n^2}(v \cup u)$.

Suppose that $G = \mathbb{F}$ is a field. By Poincaré duality,

$$\cap[M] : H^n(M;\mathbb{F}) \longrightarrow H_n(M;\mathbb{F})$$

is an isomorphism.

By the universal coefficient theorem, Theorem 5.5.19,

$$e : H_n(M;\mathbb{F}) \longrightarrow \mathrm{Hom}\,(H^n(M;\mathbb{F}),\mathbb{F}).$$

is an isomorphism. Hence the composition

$$e(\cap[M]) : H^n(M;\mathbb{F}) \longrightarrow \mathrm{Hom}\,(H^n(M;\mathbb{F}),\mathbb{F}).$$

is an isomorphism. But this composition is given by, using Theorem 5.6.13(7),

$$e(\cap[M])(v)(u) = e(u,v \cap [M]) = e(u \cup v, [M]).$$

In the language of Definition B.1.3, this shows that the map β for this form is an isomorphism, and then by Remark B.1.6 we have that the form is nonsingular.

Now consider the case $G = \mathbb{Z}$. First note that this bilinear form is well-defined on $K^n(M;\mathbb{Z}) \otimes K^n(M;\mathbb{Z})$, as if u is a torsion class, with $ru = 0$, say, and v is any class, then $0 = 0 \cup v = (ru) \cup v = r(u \cup v)$ so $u \cup v = 0$ as $H^{2n}(M;\mathbb{Z})$ is a free abelian group, and similarly for $v \cup u$.

Write $H^n(M;\mathbb{Z})$ as $F \oplus T$ when F is a free abelian group and T is the torsion subgroup. (F is in general not unique, but simply make a choice). Then under the projection $H^n(M;\mathbb{Z}) \to K^n(M;\mathbb{Z})$, F is mapped isomorphically onto K^n, and indeed this projection is an isometry between the restriction of $\langle\,,\,\rangle$ to $F \otimes F$ and $\langle\,,\,\rangle$ on $K^n(M;\mathbb{Z}) \otimes K^n(M;\mathbb{Z})$.

Now by Poincaré duality

$$\cap[M] : H^n(M;\mathbb{Z}) \longrightarrow H_n(M;\mathbb{Z})$$

is an isomorphism, so if F' is the image of F under this isomorphism, we have $H_n(M;\mathbb{Z}) = F' \oplus T'$ where T' is its torsion subgroup.

By the universal coefficient theorem, Theorem 5.5.19,

$$e : H_n(M;\mathbb{Z}) \longrightarrow \mathrm{Hom}\,(H^n(M;\mathbb{Z}),\mathbb{Z})$$

is an isomorphism, i.e.,

$$e : F' \oplus T' \longrightarrow \mathrm{Hom}\,(F \oplus T,\mathbb{Z})$$

is an isomorphism. But $\mathrm{Hom}\,(T,\mathbb{Z}) = 0$, and $\mathrm{Hom}\,(F,\mathbb{Z})$ is free, so the only map $T' \to \mathrm{Hom}\,(F,\mathbb{Z})$ is the 0 map. Thus

$$e : F' \longrightarrow \mathrm{Hom}\,(F,\mathbb{Z})$$

is an epimorphism. But this is a map between free abelian groups of the same rank, so it must be an isomorphism. Then we complete the argument exactly as in the field case.   □

Definition 6.4.9. The abelian form $\langle\,,\,\rangle$ of Theorem 6.4.8 is the *intersection form* of M.   ◊

Definition 6.4.10. Let M be a compact connected oriented manifold of dimension $2n$ for n even. The *signature* (or *index*) $\sigma(M)$ is the signature of the intersection form of M on $H^{2n}(M;\mathbb{R}) \otimes H^{2n}(M;\mathbb{R})$ as defined in Definition B.2.6.   ◊

Theorem 6.4.8 is true for arbitrary compact connected manifolds with $\mathbb{Z}/2\mathbb{Z}$ coefficients, but is not very useful. However, in the oriented case, it is extremely useful, as we now see.

Corollary 6.4.11. *Let M be a compact connected oriented manifold of dimension $2n$, n odd. Then for $G = \mathbb{Z}$ or any field $\mathbb{F}$ of characteristic not equal to 2, $\mathrm{rank}\,(K^n(M;G))$ is even. Also, the Euler characteristic $\chi(M)$ is even.*

Proof. By Theorem 6.4.8, $\langle\,,\,\rangle$ is a nonsingular skew-symmetric bilinear form on $K^n(M;G)$, so by Theorem B.2.1, $K^n(M;G)$ must have even rank.

We may use any field to compute Euler characteristic. Choosing $\mathbb{F} = \mathbb{Q}$, say, and using Poincaré duality, a short calculation shows

$$\chi(M) = \sum_{k=0}^{2n}(-1)^k \dim H_k(M;\mathbb{Q})$$

$$= 2\left(\sum_{k=0}^{n-1}(-1)^k \dim H_k(M;\mathbb{Q})\right) - \dim H_n(M;\mathbb{Q})$$

which is even.   □

Definition 6.4.12. (1) Two compact connected oriented n-manifolds M and N with fundamental classes $[M]$ and $[N]$ are of the same oriented homotopy type (resp. same oriented homeomorphism type) if there is a homotopy equivalence (resp. a homeomorphism) $f : M \to N$ with $f_*([M]) = [N]$.

(2) If M is a compact connected orientable manifold then a homotopy equivalence (or a homeomorphism) $f : M \to M$ is orientation preserving (resp. orientation reversing) if $f_* : H_n(M;\mathbb{Z}) \to H_n(M;\mathbb{Z})$ is multiplication by 1 (resp. by -1).
◊

Corollary 6.4.13. *Let M be a compact connected oriented n-manifold. The isomorphism class of the intersection form on M is an invariant of the oriented homotopy type of M.*

The finest information comes from considering the intersection form over $\mathbb{Z}$, but that may lead to difficult algebraic questions. However, the information provided by the intersection form over $\mathbb{R}$ is still enough to obtain interesting results.

Corollary 6.4.14. *Let M be a compact connected oriented manifold of dimension $2n$ with n even. If the signature $\sigma(M) \neq 0$, then there is no orientation-reversing homotopy equivalence (and hence no orientation-reversing homeomorphism) $f : M \to M$.*

Theorem 6.4.15. *Let M be a compact connected oriented manifold of dimension $2n$ with n even. If the signature $\sigma(M) \neq 0$, then M is not the boundary of an oriented $(2n+1)$-manifold.*

Proof. Suppose that M is the boundary of X^{2n+1}. Let $V = H^n(M^{2n};\mathbb{R})$, and let V have dimension t. We will use Lefschetz duality to find a subspace V_0 of V of dimension $t/2$ with the restriction of the intersection form on M to V_0 identically 0. By Lemma B.2.8, this shows that $\sigma(M) = 0$. (This also shows that if t is odd, M is not the boundary of an oriented $(2n+1)$-manifold, but this is implied by our hypothesis, as if t is odd, $\sigma(M)$ must be nonzero.)

Consider the diagram (we omit the coefficients $\mathbb{R}$)

$$
\begin{array}{ccccc}
H^n(X) & \xrightarrow{\ i^*\ } & H^n(M) & \xrightarrow{\ \delta\ } & H^{n+1}(X,M) \\
{\scriptstyle\cong}\downarrow & & {\scriptstyle\cong}\downarrow & & {\scriptstyle\cong}\downarrow \\
H_{n+1}(X,M) & \xrightarrow{\ \partial\ } & H_n(M) & \xrightarrow{\ i_*\ } & H_n(X)
\end{array}
$$

which is commutative up to sign and where the vertical maps are isomorphisms. Then

$$\dim H^n(M) = \dim H_n(M) = \dim \operatorname{Ker}(i_*) + \dim \operatorname{Im}(i_*)$$

$$= \dim \operatorname{Ker}(\delta) + \dim \operatorname{Im}(i_*)$$

$$= \dim \operatorname{Im}(i^*) + \dim \operatorname{Im}(i_*).$$

But we have the commutative diagram of Theorem 5.5.12

$$
\begin{array}{ccc}
H^n(X) & \longrightarrow & \mathrm{Hom}(H_n(X),\mathbb{R}) \\
\downarrow{\scriptstyle i^*} & & \downarrow{\scriptstyle \mathrm{Hom}(i_*,1)} \\
H^n(M) & \longrightarrow & \mathrm{Hom}(H_n(M),\mathbb{R})
\end{array}
$$

with the horizontal maps isomorphisms, and this implies that

$$\dim \mathrm{Im}\,(i^*) = \dim \mathrm{Im}\,(i_*).$$

Thus we conclude that $V_0 = \mathrm{Im}\,(i^*)$ is a subspace of $V = H^n(M)$ with $\dim V_0 = (1/2)\dim V$. Now we must investigate the cup product on V_0.

Recall that $i_*([M]) = 0 \in H_{2n}(X)$ by Corollary 6.2.36.

Now let $\alpha, \beta \in V_0$ so that $\alpha = i^*(\gamma)$ and $\beta = i^*(\delta)$ for $\gamma, \delta \in H^n(X)$. Then

$$
\begin{aligned}
\langle \alpha, \beta \rangle &= \langle \alpha \cup \beta, [M] \rangle = \langle i^*(\gamma) \cup i^*(\delta), [M] \rangle \\
&= \langle i^*(\gamma \cup \delta), [M] \rangle = \langle \gamma \cup \delta, i_*([M]) \rangle \\
&= \langle \gamma \cup \delta, 0 \rangle = 0
\end{aligned}
$$

completing the proof. $\square$

In order to give examples for these two theorems we consider the connected sum construction of oriented manifolds, a construction that is important in its own right. The basic idea is very simple, but we will have to exercise some care to ensure that the orientation comes out right.

Definition 6.4.16. Let M and N both be compact connected oriented n-manifolds, $n > 0$, with fundamental classes $[M]$ and $[N]$ respectively. Let $\varphi_\alpha : \mathbb{R}^n \to U_\alpha$ be a coordinate patch on M and $\psi_\beta : \mathbb{R}^n \to V_\beta$ be a coordinate patch on N. Let $\mathring{D}^n$ be the open unit ball in $\mathbb{R}^n$ and let S^{n-1} be the unit sphere in $\mathbb{R}^n$. Let $M' = M - \varphi_\alpha(\mathring{D}^n)$ and observe that M' is a manifold with boundary $\partial M = \varphi(S^{n-1})$ homeomorphic to S^{n-1}.

We have isomorphisms on homology

$$H_n(M) \longrightarrow H_n(M, \varphi_\alpha(\mathring{D}^n)) \longrightarrow H_n(M', \partial M')$$

where the first isomorphism comes from the inclusion of pairs $(M, \emptyset) \to (M, \varphi_\alpha(D^n))$ and the second is the inverse of excision. (Note we can apply excision here by Theorem 3.2.7.) Let $[M', \partial M']$ be the fundamental class of the manifold with boundary M' which is the image of the fundamental class $[M]$ of M under this isomorphism, and let $[\partial M'] = \partial([M', \partial M']) \in H_{n-1}(\partial M')$. Define N', $[N', \partial N']$, and $[\partial N']$ similarly.

Let $C = S^{n-1} \times [-1, 1]$. Let $i_j : S^{n-1} \to S^{n-1} \times \{j\} \subset C$ for $j = -1, 0$ or 1. Choose a fundamental class $[S^{n-1}] \in H_{n-1}(S^{n-1})$ and let $[S_0^{n-1}] = (i_0)_*([S^{n-1}]) \in H_{n-1}(C)$

and $[S_j^{n-1}] = (i_j)_*([S^{n-1}]) \in H_{n-1}(S^{n-1} \times \{j\})$ for $j = -1$ or 1. Observe that under the inclusions $S^{n-1} \times \{j\} \to C$, the image of $[S_j^{n-1}]$ is $[S_0^{n-1}]$, $j = -1, 1$. Choose the fundamental class $[C, \partial C] \in H_n(C, \partial C)$ so that $[\partial C] = \partial([C, \partial C]) = [S_1^{n-1}] - [S_{-1}^{n-1}] \in H_{n-1}(\partial C)$.

Now let $f_{-1} : S^{n-1} \times \{-1\} \to \partial M'$ be a homeomorphism with $(f_{-1})_*([S_{-1}^{n-1}]) = [\partial M'] \in H_{n-1}(\partial M')$ and let $f_1 : S^{n-1} \times \{1\} \to \partial N'$ be a homeomorphism with $(f_1)_*([S_1^{n-1}]) = -[\partial N'] \in H_{n-1}(\partial N')$. The *connected sum* $M \# N$ is the identification space

$$M \# N = M' \cup N' \cup C / \sim$$

under the identification $(s, -1) \sim f_{-1}(s)$ and $(s, 1) \sim f_1(s)$ for $s \in S^{n-1}$.

$M \# N$ is clearly a manifold and we give it the orientation with fundamental class $[M \# N]$ as described in Theorem 6.4.17 below, making it into an oriented manifold. ◇

There are many choices that went into the construction of $M \# N$, but in fact $M \# N$ is well-defined up to oriented homeomorphism type.

We summarize the (co)homological properties of the connected sum in the following theorem.

Theorem 6.4.17. *Let M and N be compact connected oriented n-manifolds, $n > 0$, with fundamental classes $[M]$ and $[N]$ respectively. Then*

$$H_n(M \# N) \cong H^n(M \# N) \cong \mathbb{Z}$$

$$\left. \begin{array}{l} H_j(M \# N) \cong H_j(M) \oplus H_j(N) \\ H^j(M \# N) \cong H^j(M) \oplus H^j(N) \end{array} \right\} \quad \text{for } 1 \le j \le n-1$$

$$H_0(M \# N) \cong H^0(M \# N) \cong \mathbb{Z}.$$

There is a fundamental class $[M \# N] \in H_n(M \# N)$ with image $[M', \partial M']$ under the isomorphism

$$H_n(M \# N) \longrightarrow H_n(M \# N, M \# N - M') \longrightarrow H_n(M', \partial M')$$

and with image $[N', \partial N']$ under the isomorphism

$$H_n(M \# N) \longrightarrow H_n(M \# N, M \# N - N') \longrightarrow H_n(N', \partial N').$$

Under the isomorphisms above, the cup product structure on $M \# N$ is as follows:

$$\text{Let } (\alpha, \gamma) \in H^j(M) \oplus H^j(N)$$

$$\text{and } (\beta, \delta) \in H^k(M) \oplus H^k(N).$$

For $j,k > 0$ and $j+k < n$,

$$(\alpha,\gamma) \cup (\beta,\delta) = (\alpha \cup \beta, \gamma \cup \delta) \in H^{j+k}(M) \oplus H^{j+k}(N).$$

For $j,k > 0$ and $j+k = n$,

$$e((\alpha,\gamma) \cup (\beta,\delta), [M\#N]) = e(\alpha \cup \beta, [M]) + e(\gamma \cup \delta, [N]),$$

i.e., if $\alpha \cup \beta = x\{M\}$ and $\gamma \cup \delta = y\{N\}$, then

$$(\alpha,\gamma) \cup (\beta,\delta) = (x+y)\{M\#N\}.$$

In particular, if $\langle \, , \, \rangle$ is the intersection form on M, $\langle \, , \, \rangle'$ the intersection form on N, and $\langle \, , \, \rangle''$ the intersection form on $M\#N$, then

$$\langle \, , \, \rangle'' = \langle \, , \, \rangle \oplus \langle \, , \, \rangle'.$$

Proof. We work in cohomology as we wish to obtain the cup product structure. The argument in homology is very similar.

Consider the disjoint union $M \cup N$ of M and N. Then it is certainly true that $H^j(M \cup N) \cong H^j(M) \oplus H^j(N)$ for every j, and that the cup product on $M \cup N$ is given by, using this direct sum decomposition,

$$(\alpha,0) \cup (\beta,0) = (\alpha \cup \beta, 0)$$
$$(0,\gamma) \cup (0,\delta) = (0, \gamma \cup \delta)$$
$$(\alpha,0) \cup (0,\delta) = (0,0)$$
$$(0,\gamma) \cup (\beta,0) = (0,0)$$

for all cohomology classes, which is the multiplication

$$(\alpha,\gamma) \cup (\beta,\delta) = (\alpha \cup \beta, \gamma \cup \delta)$$

for all cohomology classes.

Next we have the identification map $i : M \cup N \to M \vee N$, which induces an isomorphism on cohomology in all positive dimensions, and hence gives the same product structure on cohomology in positive dimensions. This is just the identity $i^*(\mu) \cup i^*(\nu) = i^*(\mu \cup \nu)$.

Consider the exact sequence of the pair (M,M'). We see, using excision,

$$\begin{array}{ccccccc} H^{j+1}(M,M') & \longleftarrow & H^j(M') & \longleftarrow & H^j(M) & \longleftarrow & H^j(M,M') \\ \Big\downarrow{\scriptstyle \cong} & & & & & & \Big\downarrow{\scriptstyle \cong} \\ H^{j+1}(D^n, S^{n-1}) & & & & & & H^j(D^n, S^{n-1}) \end{array}$$

so $H^j(M) \to H^j(M')$ is an isomorphism for $j < n-1$, and an injection for $j = n-1$. But we also have the exact sequence

$$H^n(M') \longleftarrow H^n(M) \longleftarrow H^n(M,M') \longleftarrow H^{n-1}(M') \longleftarrow H^{n-1}(M)$$

We have that $H^n(M) \cong \mathbb{Z}$ and $H^n(M,M') \cong \mathbb{Z}$. But M' is a manifold with boundary, so by Lefschetz duality $H^n(M') \cong H_0(M', \partial M') = 0$ as the map $H_0(\partial M') \to H_0(M')$ is surjective (in fact, an isomorphism).

Hence the map $H^n(M,M') \to H^n(M)$ is an isomorphism, and in particular is an injection, so $H^j(M) \to H^j(M')$ is also a surjection for $j = n-1$, and hence an isomorphism.

Of course, the map $H^j(N) \to H^j(N')$ is an isomorphism for all $j \le n-1$ by exactly the same argument.

Let $M'' = M' \cup S^{n-1} \times [-1,0] \subset M\#N$ and let $N'' = N' \cup S^{n-1} \times [0,1] \subset M\#N$. Then M' is a strong deformation retract of M'' and N' is a strong deformation retract of N'', so the inclusions/retractions induce isomorphisms on cohomology. (In fact, M'' is homeomorphic to M' and N'' is homeomorphic to N'.) Also, $M\#N = M'' \cup N''$ and $M'' \cap N'' = S^{n-1} \times \{0\}$.

Finally, observe that there is a natural map $p : M\#N \to M \vee N$ obtained by identifying $S^{n-1} \times \{0\}$ to the point $q = M \cap N \in M \vee N$.

Here is a picture of the various spaces involved.

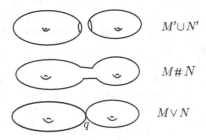

$$M' \cup N'$$
$$M\#N$$
$$M \vee N$$

With all these preliminaries out of the way, we have a commutative diagram of Mayer-Vietoris sequences

$$H^j(q) \longleftarrow H^j(M) \oplus H^j(N) \longleftarrow H^j(M \cup N) \longleftarrow H^{j-1}(q) \longleftarrow H^{j-1}(M) \oplus H^{j-1}(N)$$
$$\downarrow \qquad\qquad \downarrow \qquad\qquad \downarrow \qquad\qquad \downarrow \qquad\qquad \downarrow$$
$$H^j(S^{n-1} \times \{0\}) \longleftarrow H^j(M'') \oplus H^j(N'') \longleftarrow H^j(M\#N) \longleftarrow H^{j-1}(S^{n-1} \times \{0\}) \longleftarrow H^{j-1}(M'') \oplus H^{j-1}(N'')$$

For $j \ge 1$ and $j < n-1$, the first, second, fourth, and fifth vertical arrows are isomorphisms, and hence, by the five lemma, Lemma A.1.8, so is the third.

In case $j = n-1$, consider the map in the lower left-hand corner. Note that $S^{n-1} \times \{0\}$ is the boundary of both M'' and N'', so the maps $H^{n-1}(M'') \to H^{n-1}(S^{n-1} \times \{0\})$ and $H^{n-1}(N'') \to H^{n-1}(S^{n-1} \times \{0\})$ are both the 0 map by the dual of Corollary 6.2.36. Hence we may replace $H^{n-1}(S^{n-1} \times \{0\})$ by the zero group and apply the five lemma once again.

Now consider the case $j = n$. We have the lower sequence

$$\longleftarrow H^n(M\#N) \longleftarrow H^{n-1}(S^{n-1} \times \{0\}) \longleftarrow H^{n-1}(M'') \oplus H^{n-1}(N'')$$

$$H^n(M'') \oplus H^n(N'')$$

We have just observed that the right-hand map is the zero map. Also, we observed earlier that $H^n(M'') = 0$ and $H^n(N'') = 0$ and hence $H^n(M'') = 0$ and $H^n(N'') = 0$. Thus $H^n(M\#N) \cong \mathbb{Z}$.

Let $\{M\}$ denote the fundamental cohomology class of M. Consider the sequence

$$H^n(M) \xleftarrow{\cong} H^n(M,D) \xrightarrow{\cong} H^n(M',\partial M') \xleftarrow{\cong} H^n(M\#N,N'') \xrightarrow{\cong} H^n(M\#N).$$

The first map is certainly an isomorphism, the next two are excisions, and the last map is a map in the exact sequence of the pair $(M\#N,N'')$ whose next term is $H^n(N'') = 0$. Thus it is a surjection from $\mathbb{Z}$ to $\mathbb{Z}$ and hence an isomorphism. The image of $\{M\}$ in the intermediate group is the fundamental cohomology class $\{M',\partial M'\}$ and its image in the right-hand group is the fundamental cohomology class $\{M\#N\}$. Similarly the image of $\{N\}$ is $\{M\#N\}$. (These follow because of our choices of orientations.)

Now we have the same argument that the cup product structures on $H^*(M \vee N)$ and $H^*(M\#N)$ are isomorphic for cohomology classes of dimensions j and k with $j,k > 0$ and $j+k < n$. If $\mu \in H^j(M\#N)$ and $v \in H^k(M\#N)$, then $p^* : H^i(M\#N) \to H^i(M \vee N)$ is an isomorphism for $i = j,k,j+k$, and $p^*(\mu) \cup p^*(v) = p^*(\mu \cup v)$.

If $j + k = n$, consider a class $p^*(\alpha,0)$ in $H^j(M\#N)$ and a class $p^*(0,\gamma)$ in $H^k(M\#N)$. Then $p^*(\alpha,0) \cup p^*(0,\gamma) = p^*((\alpha,0) \cup (0,\gamma)) = p^*(0) = 0$ as $(\alpha,0) \cup (0,\gamma) = 0 \in H^{j+k}(M \vee N)$. Similarly $p^*(0,\beta) \cup p^*(\gamma,0) = 0$.

Again with $j+k = n$, consider a class $p^*(\alpha,0) \in H^j(M\#N)$ and a class $p^*(\beta,0) \in H^k(M\#N)$. Suppose $\alpha \cup \beta = x\{M\} \in H^n(M)$. Then $p^*(\alpha,0) \cup p^*(\beta,0) = p^*(\alpha \cup \beta,0) = p^*(x\{M\},0) = x\{M+N\}$. Similarly for $p^*(0,\gamma) \in H^j(M\#N)$ and $p^*(0,\delta) \in H^k(M\#N)$ with $\gamma \cup \delta = y\{N\}$ we have $p^*(0,\gamma) \cup p^*(0,\delta) = y\{M+N\}$, and we are done.   □

Example 6.4.18. Let n be even. We adopt the notation and language of Example 6.4.4.

In the basis of $H^n(\mathbb{C}P^n)$ consisting of the element $\alpha^{n/2}$, intersection form of the oriented manifold $\mathbb{C}P^n$ has matrix $[1]$, so $H(\mathbb{C}P^n)$ has rank 1 and signature $\sigma(\mathbb{C}P^n) = 1$, while the intersection form of the oriented manifold $\overline{\mathbb{C}P^n}$ has matrix $[-1]$, so $\overline{\mathbb{C}P^n}$ has signature $\sigma(\overline{\mathbb{C}P^n}) = -1$.

Then for nonnegative integers r and s,

$$M = \mathbb{C}P^n \# \cdots \# \mathbb{C}P^n \# \overline{\mathbb{C}P^n} \# \cdots \# \overline{\mathbb{C}P^n},$$

where there are r copies of $\mathbb{C}P^n$ and s copies of $\overline{\mathbb{C}P^n}$, is a $(2n)$-manifold with rank $H_n(M) = r + s$ and signature $\sigma(M) = r - s$.   ◊

6.5 Exercises

Exercise 6.5.1. (a) A manifold is defined to be a topological space (i) in which every point has a neighborhood homeomorphic to $\mathbb{R}^n$, for some fixed n, (ii) which is Hausdorff, and (iii) which is separable. Give examples of spaces satisfying any two of these properties but not the third.
(b) Show that if M is compact, then (i) implies (iii).

Exercise 6.5.2. Let $f : (M, \partial M) \to (N, \partial N)$ be a homeomorphism. Show that $f : \text{int}(M) \to \text{int}(N)$ and $f : \partial M \to \partial N$.

Exercise 6.5.3. Show that every connected manifold is homogeneous, i.e., that if M is a connected manifold and x and y are any two pairs of M, there is a homeomorphism $f : M \to M$ with $f(x) = y$. (Note this implies that if M and N are connected manifolds, $M \vee N$ is well-defined up to homeomorphism.)

Exercise 6.5.4. Let M and N be manifolds of dimension $n \geq 1$, and let $M \vee N$ be formed from M and N by identifying $x \in M$ with $y \in N$. Call this identification point z. Suppose that $f : M \vee N \to M \vee N$ is any homeomorphism. Show that $f(z) = z$.

Exercise 6.5.5. Let M and N be connected manifolds. Describe $H_*(M \vee N)$, $H^*(M \vee N)$, and the ring structure on $H^*(M \vee N)$, in terms of $H_*(M), H_*(N), H^*(M)$, $H^*(N)$, and the ring structures on $H^*(M)$ and $H^*(N)$.

Exercise 6.5.6. Prove Corollary 6.2.31.

Exercise 6.5.7. Let M and N be manifolds. Show that $M \times N$ is orientable if and only if both M and N are orientable.

Exercise 6.5.8. Let M and N be manifolds. Express the first Stiefel-Whitney class $w_1(M \times N)$ in terms of $w_1(M)$ and $w_1(N)$.

Exercise 6.5.9. Let M be an n-manifold and suppose $H_*(M) \neq H_*(S^n)$. Show that the suspension ΣM of M is not a manifold. (Of course, ΣS^n is a manifold as it is homeomorphic to S^{n+1}.)

Exercise 6.5.10. Let $f : X \to X$. The mapping torus T_f of f is the quotient space $X \times I / \sim$ where $\sim$ is the identification of $(f(x), 1)$ with $(x, 0)$ for every $x \in X$. If X is an n-manifold and f is a homeomorphism, show that T_f is an $(n+1)$-manifold.

Exercise 6.5.11. Let $f : S^n \to S^n$ be a homeomorphism. Find $H_*(T_f), H^*(T_f)$, and the ring structure on $H^*(T_f)$.

Exercise 6.5.12. Let $f : M \to M$ be a homeomorphism, where M is an n-manifold.

(a) If M is nonorientable, show that T_f is nonorientable.
(b) If M is oriented, show that T_f is orientable if and only if f is orientation-preserving (i.e., if $f_*([M]) = [M]$).

Exercise 6.5.13. Let M and N be connected compact oriented n-manifolds, and let $f : M \to N$. Then $f_*([M]) = d[N]$ for some integer d. This integer is called the degree of f. If f has nonzero degree, show that $N = f(M)$ (i.e., that for every $y_0 \in N$, there is an $x_0 \in M$ with $f(x_0) = y_0$).

Exercise 6.5.14. Let M and N be compact connected manifolds of dimension n and let $f : M \to N$. Suppose there is some point $y_0 \in N$ such that $f^{-1}(y_0) = \{x_0\}$, a single point of M. Suppose furthermore that y_0 has an open neighborhood V such that, if $U = f^{-1}(V)$, $f : U \to V$ is a homeomorphism.

(a) Show that $f_* : H_n(M; \mathbb{Z}_2) \to H_n(N; \mathbb{Z}_2)$ is an isomorphism.
(b) If M and N are both oriented, show that f has degree ± 1.

Exercise 6.5.15. (a) Let M be a compact connected n-dimensional manifold. Show there is a map $f : M \to S^n$ with $f_* : H_n(M; \mathbb{Z}_2) \to H_n(S^n; \mathbb{Z}_2)$ an isomorphism.
(b) Let M be a compact connected oriented n-manifold. Let S^n be oriented. Show there is a degree 1 map $f : M^n \to S^n$.

Exercise 6.5.16. (a) Let M and N be compact connected n-manifolds and let $f : M \to N$ be a map with $f_* : H_n(M; \mathbb{Z}_2) \to H_n(N; \mathbb{Z}_2)$ an isomorphism. Show that $f_* : H_k(M; \mathbb{Z}_2) \to H_k(N; \mathbb{Z}_2)$ is onto for every k, and $f^* : H^k(N; \mathbb{Z}_2) \to H^k(M; \mathbb{Z}_2)$ is $1 - 1$ for every k.
(b) Let M and N be compact connected oriented n-manifolds and let $f : M \to N$ be any map of degree 1. Show that $f_* : H_k(M; \mathbb{Z}) \to H_n(N; \mathbb{Z})$ is onto for every k, and $f^* : H^k(N; \mathbb{Z}) \to H^k(M; \mathbb{Z})$ is $1 - 1$ for every k.
(c) Let M and N be compact connected oriented n-manifolds and let $f : M \to N$ be a map of nonzero degree d. Let $\mathbb{F}$ be any field of characteristic 0 or characteristic relatively prime to d. Show that $f_* : H_k(M; \mathbb{F}) \to H_k(N; \mathbb{F})$ is onto for every k, and $f^* : H^k(N; \mathbb{F}) \to H^k(M; \mathbb{F})$ is $1 - 1$ for every k.

Chapter 7
Homotopy Theory

Let X be a space and let x_0 be a point in X. In section 2.1 we introduced the fundamental group $\pi_1(X,x_0)$. In this chapter we introduce the *homotopy groups* $\pi_n(X,x_0)$ for every $n \geq 0$. Also, if (X,A,x_0) is a triple, i.e., if A is a subspace of X and x_0 is a point in A, we have the *relative homotopy groups* $\pi_n(X,A,x_0)$ for every $n \geq 1$.

Actually, what we have said is not quite precise, as $\pi_0(X)$ and $\pi_1(X,A,x_0)$ do not have a group structure. Instead they are pointed sets. Thus we digress a bit to discuss pointed sets.

A *pointed set* S is a nonempty set with a distinguished element $s_0 \in S$.

If S and T are pointed sets with distinguished elements s_0 and t_0 respectively, then a map of pointed sets $f : S \to T$ is a function $f : S \to T$ with $f(s_0) = t_0$.

In this situation,

$$\mathrm{Ker}\,(f) = \{s \in S \mid f(s) = t_0\}$$
$$\mathrm{Im}\,(f) = \{t \in T \mid t = f(s) \text{ for some } s \in S\}.$$

Our language and some of the details of our constructions in this chapter will differ here than in Chap. 2, but we leave it to the reader to check that for $n = 1$, what we are doing here agrees with what we did there.

7.1 Definitions and Basic Properties

We first establish some conventions and notation we will use throughout this chapter:

(X,x_0) always denotes a nonempty space X and a point $x_0 \in X$, and X_0 denotes the path component of X containing x_0.

(X,A,x_0) always denotes a nonempty space X, a nonempty subspace A of X, and a point $x_0 \in A$.

© Springer International Publishing Switzerland 2014
S.H. Weintraub, *Fundamentals of Algebraic Topology*, Graduate Texts
in Mathematics 270, DOI 10.1007/978-1-4939-1844-7_7

I^n is the unit cube in $\mathbb{R}^n$, $I^n = \{(t_1,\ldots,t_n) \mid 0 \leq t_i \leq 1, \, i = 1,\ldots,n\}$. Its boundary $\partial I^n = \{(t_1,\ldots,t_n) \mid t_i = 0 \text{ or } 1 \text{ for some } i\}$, and $\partial I^n = J^{n-1} \cup K^{n-1}$ where $J^{n-1} = \{(t_1,\ldots,t_n) \mid t_1 = 1\}$ and K^{n-1} is the closure of the complement of J^{n-1} in ∂I^n. We call J^{n-1} the *front* of ∂I^n and K^{n-1} the *rear*.

(Note that $\partial I^1 = \{0\} \cup \{1\}$ with $J^0 = \{1\}$ and $K^0 = \{0\}$.)

Definition 7.1.1. For $n = 0$, $\pi_n(X,x_0)$ is the set of homotopy classes of maps $\alpha : (\partial I^1, K^0) \to (X,x_0)$. For $n \geq 1$, $\pi_n(X,x_0)$ is the set of homotopy classes of maps $\alpha : (I^n, \partial I^n) \to (X,x_0)$.

For $n \geq 1$, $\pi_n(X,A,x_0)$ is the set of homotopy classes of maps $\alpha : (I^n, J^{n-1}, K^{n-1}) \to (X,A,x_0)$.   ◇

Analogously with our construction of the fundamental group, we may identify $(I^n/\partial I^n, \partial I^n/\partial I^n)$ with $(S^n, 1)$ and under this identification we see that $\pi_n(X,x_0)$ is the set of homotopy classes of maps $f : (S^n, 1) \to (X,x_0)$, for $n \geq 1$.

Under this identification f represents the trivial element of $\pi_n(X,x_0)$ if f extends to $\bar{f} : (D^{n+1}, 1) \to (X,x_0)$. However, note that the "obvious" homotopy to the constant map provided by this extension is not the correct one, as it is not a homotopy rel $\{1\}$. The correct homotopy is $F : (D^{n+1}, 1) \times I \to (X,x_0)$ given as follows: Let $i_t : S^n \to D^{n+1}$ by $i_t(x_1,\ldots,x_{n+1}) = (1-t)(x_1,\ldots,x_{n+1}) + (t,0,\ldots,0)$. (The image of S^n under i_t is a sphere of radius $1-t$ centered at the point $(t,0,\ldots,0)$. Also, $i_t(1,0,\ldots,0) = (1,0,\ldots,0)$ for every t.) Then $F(x_1,\ldots,x_{n+1},t) = \bar{f}(i_t(x_1,\ldots,x_{n+1}))$.

Similarly we may regard $\pi_n(X,A,x_0)$ as the group of homotopy classes of maps $f : (D^n, S^{n-1}, 1) \to (X,A,x_0)$ for $n \geq 2$.

We observe that $\pi_0(X,x_0)$ and $\pi_1(X,A,x_0)$ are pointed sets with distinguished element the homotopy class of the constant map to x_0.

Lemma 7.1.2. $\pi_0(X,x_0)$ *is isomorphic to the set of path components of X, with distinguished element X_0.*

We now put a group structure on $\pi_n(X,x_0)$ for $n \geq 1$ and on $\pi_n(X,A,x_0)$ for $n \geq 2$. In the following pictures heavy points, lines, or regions mark points that map to the basepoint x_0.

Definition 7.1.3. The group structure on $\pi_n(X,x_0)$ for $n \geq 1$ and on $\pi_n(X,A,x_0)$ for $n \geq 2$ is given by "following":

For $n = 1$ the product of

For $n = 2$ the composition of

(In the case of $\pi_2(X,x_0)$ the front face is also heavy, while in the case of $\pi_2(X,A,x_0)$ it is not.)

Similarly for $n > 2$.   ◊

Lemma 7.1.4. *For $n \geq 2$, $\pi_n(X,x_0)$ is an abelian group. For $n \geq 3$, $\pi_n(X,A,x_0)$ is an abelian group.*

Proof. Here is a picture of a homotopy between $\alpha\beta$ and $\beta\alpha$ in case $n = 2$, for $\pi_n(X,x_0)$.

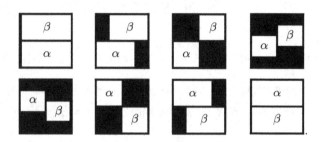

Similarly for $n > 2$ for $\pi_n(X,x_0)$ and for $n \geq 3$ for $\pi_n(X,A,x_0)$.   □

Definition 7.1.5. The boundary map $\partial : \pi_n(X,A,x_0) \to \pi_{n-1}(A,x_0)$ is defined as follows: Let $\alpha : (I^n, J^{n-1}, K^{n-1}) \to (X,A,x_0)$ represent an element of $\pi_n(X,A,x_0)$. If $n = 1$, let $i : \partial I^1 \to I^1$ be the inclusion, and observe that $i(K_0) = K_0$. Then $\partial\alpha$ is the homotopy class of the map $\alpha i : (\partial I^1, K^0) \to (A,x_0)$. If $n \geq 2$, let $i : I^{n-1} \to I^n$ by $i(t_1,\ldots,t_{n-1}) = (1,t_1,\ldots,t_{n-1})$, and observe that $i(I^{n-1}) = J^{n-1}$ and $i(\partial I^{n-1}) \subseteq K^{n-1}$. Then $\partial\alpha$ is the homotopy class of the map $\alpha i : (I^{n-1}, \partial I^{n-1}) \to (A,x_0)$.   ◊

Definition 7.1.6. Let $f : (X,x_0) \to (Y,y_0)$ be a map. The map f induces $f_* : \pi_n(X,x_0) \to \pi_n(Y,y_0)$ by composition, i.e., if $\alpha : (\partial I^1, K^0) \to (X,x_0)$ represents an element of $\pi_0(X,x_0)$, or $\alpha : (I^n, \partial I^n) \to (X,x_0)$ represents an element of $\pi_n(X,x_0)$ for $n \geq 1$, then $f_*(\alpha)$ is the homotopy class of the map $f\alpha$. Similarly a map $f : (X,A,x_0) \to (Y,B,y_0)$ induces $f_* : \pi_n(X,A,x_0) \to \pi_n(Y,B,y_0)$.   ◊

Lemma 7.1.7. *For $n \geq 1$, $f_* : \pi_n(X,x_0) \to \pi_n(Y,y_0)$ is a group homomorphism, and for $n \geq 2$, $f_* : \pi_n(X,A,x_0) \to \pi_n(Y,B,y_0)$ is a group homomorphism. Also, $f_* : \pi_0(X,x_0) \to \pi_0(Y,y_0)$ and $f_* : \pi_1(X,A,x_0) \to \pi_1(Y,B,y_0)$ are maps of pointed sets.*

Lemma 7.1.8. *The inclusion $(X_0,x_0) \to (X,x_0)$ induces isomorphisms $\pi_n(X_0,x_0) \to \pi_n(X_0,x_0)$ for every $n \geq 1$, and isomorphisms $\pi_n(X_0, X_0 \cap A, x_0) \to \pi_n(X,A,x_0)$ for every $n \geq 1$.*

We have the following basic properties of homotopy groups (compare the Eilenberg-Steenrod axioms for homology): In parts (1), (2), and (5) we only state these for relative homotopy groups, but this also includes the absolute case, as $\pi_n(X,x_0,x_0) = \pi_n(X,x_0)$ for $n \geq 1$, and in this special case we define $\pi_0(X,x_0,x_0)$ to be $\pi_0(X,x_0)$.

Theorem 7.1.9. (1) *If $f : (X,A,x_0) \to (X,A,x_0)$ is the identity map, then $f_* : \pi_n(X,A,x_0) \to \pi_n(X,A,x_0)$ is the identity map.*

(2) *If $f : (X,A,x_0) \to (Y,B,y_0)$ and $g : (Y,B,y_0) \to (Z,C,z_0)$, and h is the composition $h = g \circ f$, $h : (X,A,x_0) \to (Z,C,z_0)$, then $h_* = g_* \circ f_*$.*

(3) *If $f : (X,A,x_0) \to (Y,B,y_0)$ then the following diagram commutes:*

$$
\begin{array}{ccc}
\pi_n(X,A,x_0) & \xrightarrow{\ f_*\ } & \pi_n(Y,B,y_0) \\
\ \downarrow{\partial} & & \ \downarrow{\partial} \\
\pi_{n-1}(A,x_0) & \xrightarrow{\ (f|A)_*\ } & \pi_{n-1}(B,y_0)
\end{array}
$$

(4) *The homotopy sequence*

$$\cdots \longrightarrow \pi_n(A,x_0) \longrightarrow \pi_n(X,x_0) \longrightarrow \pi_n(X,A,x_0) \longrightarrow \pi_{n-1}(A,x_0)$$

$$\longrightarrow \cdots \longrightarrow \pi_1(X,A,x_0) \longrightarrow \pi_0(A,x_0) \longrightarrow \pi_0(X,x_0)$$

is exact.

(5) *If $f : (X,A,x_0) \to (Y,B,y_0)$ and $g : (X,A,x_0) \to (Y,B,y_0)$ are homotopic, then $f_* : \pi_n(X,A,x_0) \to \pi_n(Y,B,y_0)$ and $g_* : \pi_n(X,A,x_0) \to \pi_n(Y,B,y_0)$ are equal.*

Proof. Parts (1), (2), (3), and (5) are immediate. We leave the proof of (4) as an exercise. □

Remark 7.1.10. Note there is no analog of the excision property for homotopy. It is this property that makes homology groups (relatively) easy to compute. By contrast, homotopy groups are usually (very) hard to compute. ◇

Just as in the case of the fundamental group, we can ask about what happens when we change base points, and the answer is very similar.

Theorem 7.1.11. *Let X be a path-connected space and let $x_0, x_1 \in X$. Let $\alpha : I \to X$ be a path from x_0 to x_1, i.e., $\alpha(0) = x_0$ and $\alpha(1) = x_1$. Then α induces an isomorphism $\alpha_* : \pi_n(X,x_0) \to \pi_n(X,x_1)$. Also, if α and $\beta : I \to X$ are homotopic rel ∂I, then $\alpha_* = \beta_*$.*

Proof. Let $f : (I^n, \partial I^n) \to (X,x_0)$ represent an element of $\pi_n(X,x_0)$. The following picture shows how to obtain $\alpha_*(f) \in \pi_n(X,x_1)$ for $n = 2$, with the general case being similar.

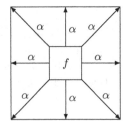

By this picture we mean we follow the path α on the portion of "rays" emanating from the center between the inner and outer boxes. □

There is no reason to restrict ourselves to $x_1 \neq x_0$. In the case $x_1 = x_0$, we have the following.

Corollary 7.1.12. *Let X be a path-connected space and let $x_0 \in X$. Then the construction of Theorem 7.1.11 gives an action of $\pi_1(X,x_0)$ on $\pi_n(X,x_0)$ for every n. The set of equivalence classes of elements of $\pi_n(X,x_0)$ under this action is in 1–1 correspondence with the set of free homotopy classes of maps $f : S^n \to X$.*

Proof. The only point to note is that we are considering homotopies $F : I^n \times I \to X$ with the property that for every $t \in I$, $F | \partial I^n \times \{t\}$ is a map to a single point, so this gives us homotopies of maps $f : S^n \to X$ where the point 1 is allowed to move during the homotopy. □

Remark 7.1.13. Note this generalizes the case $n = 1$, where the action of $\pi_1(X,x_0)$ on $\pi_1(X,x_0)$ is by conjugation. (Compare Lemma 2.1.3.) ◇

Definition 7.1.14. The space X is *n-simple* if the action of $\pi_1(X,x_0)$ on $\pi_n(X,x_0)$ is trivial. ◇

Remark 7.1.15. If X is simply-connected, then X is n-simple for every n. For example, this is the case for $X = S^m$ for every $m \geq 2$. ◇

Remark 7.1.16. If X is n-simple, then there is a canonical isomorphism from $\pi_n(X,x_0)$ to $\pi_n(X,x_1)$: Choose α_* for any path α from x_0 to x_1. Because of this, in dealing with n-simple spaces we often drop the basepoint and write $\pi_n(X)$ instead of $\pi_n(X,x_0)$. ◇

We have the following result about the homotopy of CW complexes.

Lemma 7.1.17. *Let X be a CW complex, x_0 a 0-cell in X, and let X^n be the n-skeleton of X, for any n. Let $i : X^n \to X$ be the inclusion. Then $i_* : \pi_k(X^n,x_0) \to \pi_k(X,x_0)$ is an isomorphism for $k < n$ and an epimorphism for $k = n$.*

7.2 Further Results

In this section we give further results on homotopy groups. These fit in naturally with, and extend, work we have already done. Towards the beginning of this section we give proofs that are complete, or nearly so. But, as we have said, homotopy theory is relatively hard. We cite some classical results later in this section for the edification of the reader, but their proofs are beyond the scope of this book.

Theorem 7.2.1. *Let X and Y be path-connected spaces. Let $x_0 \in X$ and $y_0 \in Y$. Let $p : X \times Y \to X$ and $q : X \times Y \to Y$ be projection on the first and second* *factors respectively. Then* $p_* \times q_* : \pi_n(X \times Y, (x_0, y_0)) \to \pi_n(X, x_0) \times \pi_n(Y, y_0)$ *is an isomorphism for all n.*

Proof. First we show $p_* \times q_*$ is onto. Let $f : (S^n, 1) \to (X, x_0)$ represent an element α of $\pi_n(X, x_0)$ and let $g : (S^n, 1) \to (Y, y_0)$ represent an element β of $\pi_n(Y, y_0)$. Then $h : (S^n, 1) \to (X \times Y, (x_0, y_0))$ by $h(t) = (f(t), g(t))$ represents an element γ of $\pi_n(X \times Y, (x_0, y_0))$ with $p_*(\gamma) = \alpha$ and $q_*(\gamma) = \beta$.

Next we show $p_* \times q_*$ is 1-1: Let f and g be as above and suppose h represents the trivial element of $\pi_n(X \times Y, (x_0, y_0))$. Then h extends to a map $H : (S^n, 1) \times I \to (X \times Y, (x_0, y_0))$ with $H|(S^n, 1) \times \{1\}$ the constant map to the point (x_0, y_0). Then $F = p_*(H)$ is an extension of f with $F|(S^n, 1) \times \{1\}$ the constant map to the point x_0, and $G = q_*(H)$ is an extension of g with $G|(S^n, 1) \times \{1\}$ the constant map to the point y_0. □

Recall we introduced the notion of a covering space in Sect. 2.2.

Theorem 7.2.2. *Let X be a path-connected space and let $\tilde{X}$ be a connected covering space of X. Let $p : \tilde{X} \to X$ be the covering projection. Let $\tilde{x}_0 \in \tilde{X}$ and let $x_0 \in X$ with $p(\tilde{x}_0) = x_0$. Then for every $n \geq 2$, $p_* : \pi_n(\tilde{X}, \tilde{x}_0) \to \pi_n(X, x_0)$ is an isomorphism.*

Proof. First we show p_* is onto. Let $f : (S^n, 1) \to (X, x_0)$ represent an element of $\pi_n(X, x_0)$. We wish to show there is an $\tilde{f} : (S^n, 1) \to (\tilde{X}, \tilde{x}_0)$ making the following diagram commute

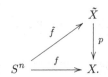

But, by Corollary 2.3.3, $\pi_1(S^n) = 0$, so by Theorem 2.2.8, such as $\tilde{f}$ always exists.

Next we show p_* is 1–1. Let $\tilde{f} : (S^n, 1) \to (\tilde{X}, \tilde{x}_0)$ and $f = p\tilde{f} : (S^n, 1) \to (X, x_0)$. Suppose that f represents the trivial element of $\pi_n(X, x_0)$, so that f extends to $F : (S^n, 1) \times I \to (X, x_0)$ with $F|(S^n, 1) \times \{1\}$ the constant map to the point x_0. Then, by Theorem 2.2.5, there is a map $\tilde{F} : (S^n, 1) \times I \to (\tilde{X}, \tilde{x}_0)$ with $p\tilde{F} = F$. In particular, $\tilde{F}|(S^n, 1) \times \{1\}$ has image contained in $p^{-1}(x_0)$. But $p^{-1}(x_0)$ is a set of discrete points, and S^n is connected, so $\tilde{F}|(S^n, 1) \times \{1\}$ has image a single point. But $\tilde{F}|(1, 1) = \tilde{x}_0$ so that point is $\tilde{x}_0$, and hence $\tilde{f}$ represents the trivial element of $\pi_n(X, x_0)$. □

Corollary 7.2.3. *For every $n \geq 2$, $\pi_n(S^1, 1) = 0$.*

Proof. We know from Example 2.2.3 that $p : \mathbb{R} \to S^1$ by $p(t) = \exp(2\pi i t)$ is a covering map, so for $n \geq 2$, $\pi_n(S^1, 1) \cong \pi_n(\mathbb{R}, 0) = 0$ as $\mathbb{R}$ is contractible. □

The proof used the covering homotopy property, which for the convenience of the reader we restate here.

Definition 7.2.4. A map $p : Y \to X$ has the *covering homotopy property* if the following holds: Let E be an arbitrary space and let $F : E \times I \to X$ be an arbitrary map. Let $\tilde{f} : E \times \{0\} \to Y$ be an arbitrary map such that $p\tilde{f}(e,0) = F(e,0)$ for every $e \in E$. Then $\tilde{f}$ extends to a map $\tilde{F} : E \times I \to Y$ with $F = p\tilde{F}$, i.e., to a map making the following diagram commute

$$
\begin{array}{ccc}
E \times \{0\} & \xrightarrow{\tilde{f}} & Y \\
\downarrow & \nearrow{\tilde{F}} & \downarrow{p} \\
E \times I & \xrightarrow{F} & X.
\end{array}
$$

If $p : Y \to X$ has the covering homotopy property then p is a *fibration*. ◊

Example 7.2.5. (1) If X and Y are any two spaces, the projection $p : X \times Y \to X$ onto the first factor is a fibration.
(2) A covering projection $p : \tilde{X} \to X$ is a fibration. ◊

Here is one of the most important ways fibrations arise.

Definition 7.2.6. A map $p : E \to B$ is a locally trivial *fiber bundle* with fiber F if the following holds: There is a cover of E by open sets $\{U_i\}$ and for each i a homeomorphism $h_i : p^{-1}(U_i) \to U_i \times F$ making the following diagram commute

$$
\begin{array}{ccc}
p^{-1}(U_i) & \xrightarrow{h_i} & U_i \times F \\
\downarrow{p} & & \downarrow \\
U_i & \xrightarrow{} & U_i
\end{array}
$$

where the lower horizontal map is the identity and the right-hand vertical map is projection onto the first factor. The map p has a *section* s if there is a map $s : B \to E$ with $ps : B \to B$ the identity. ◊

Example 7.2.7. (1) The projection $p : X \times Y \to X$ is a locally trivial fiber bundle with fiber Y. (Indeed, we call this a globally trivial fiber bundle.)
(2) Every covering projection $p : \tilde{X} \to X$ is a locally trivial fiber bundle with fiber the discrete space $p^{-1}(x_0)$. For example, $p : S^n \to \mathbb{R}P^n$ is a fiber bundle with fiber two points.
(3) The map $p : S^{2n+1} \to \mathbb{C}P^n$ given by $p(z_0, \ldots, z_n) = [z_0, \ldots, z_n]$ is a locally trivial fiber bundle with fiber $S^1 = \{z \in \mathbb{C} \mid |z| = 1\}$. ◊

Theorem 7.2.8. *Every locally trivial fiber bundle is a fibration.*

Theorem 7.2.9. *Let $p : E \to B$ be a locally trivial fiber bundle. Let $b_0 \in B$, $F = p^{-1}(b_0)$, and $e_0 \in F$. Then for every n,*

$$p_* : \pi_n(E, F, e_0) \longrightarrow \pi_n(B, b_0)$$

is an isomorphism.

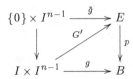

Proof. First we show p_* is onto. Let $g : (I^n, \partial I^n) \to (B, b_0)$ represent an element of $\pi_n(B, b_0)$. Regard I^n as $I \times I^{n-1}$. Then $\{0\} \times I^{n-1} \subset \partial I^{n-1}$, so we let $\tilde{g} : \{0\} \times I^{n-1} \to e_0$. Then we can apply the covering homotopy property to obtain a commutative diagram and hence a map $G' : (I^n, J', K') \to (E, F, e_0)$, where $J' = \{0\} \times I^{n-1}$ and K' is the closure of $\partial I^n - J'$. This is almost but not quite what we need to obtain a representation of an element of $\pi_n(E, F, e_0)$. For that we need $G : (I^n, J^{n-1}, K^{n-1}) \to (E, F, e_0)$. But we may obtain such a map G by composing a homeomorphism of I^n with itself with the map G' as in the following picture, where (as usual) the heavy lines indicate points mapped to e_0:

Next we show p_* is 1–1. Let $\tilde{g} : (I^n, J^{n-1}, K^{n-1}) \to (E, F, e_0)$ and suppose that $g = p_*(\tilde{g})$ represents the trivial element of $\pi_n(B, b_0)$. Then there is a mapping $G : (I^n, J^{n-1}, K^{n-1}) \times I \to (B, b_0)$ extending the map g on $(I^n, J^{n-1}, K^{n-1}) \times \{0\}$ and with $G : (I^n, J^{n-1}, K^{n-1}) \times \{1\} \to b_0$. By the covering homotopy property, there is a map $\tilde{G} : (I^n, J^{n-1}, K^{n-1}) \times I \to E$ extending $\tilde{g}$ with $p\tilde{G} = G$. In particular, $p\tilde{G}((I^n, J^{n-1}, K^{n-1}) \times \{1\}) = b_0$, i.e., $\tilde{G}(I^n \times \{1\}) \subseteq F$. But this means that $\tilde{G}$, and hence $\tilde{g}$, represents the trivial element of $\pi_n(E, F, e_0)$. □

We now have a corollary that generalizes both Theorems 7.2.1 and 7.2.2.

Corollary 7.2.10. *Let $p : E \to B$ be a locally trivial fiber bundle with fiber $F = p^{-1}(b_0)$ and let $f_0 \in F$. Then there is an exact sequence*

$$\cdots \longrightarrow \pi_n(F, f_0) \longrightarrow \pi_n(E, f_0) \xrightarrow{p_*} \pi_n(B, b_0) \longrightarrow \pi_{n-1}(F, f_0) \longrightarrow \cdots.$$

If p has a section, then for each n,

$$\pi_n(E, f_0) \cong \pi_n(B, b_0) \times \pi_n(F, f_0).$$

Proof. The first claim follows immediately from Theorems 7.1.9 and 7.2.9.

As for the second claim, if s is a section, then $s_* : \pi_n(B, b_0) \to \pi_n(E, f_0)$ splits p_*, so this long exact sequence breaks up into a series of split short exact sequences. □

Example 7.2.11. By Example 7.2.7 and Corollary 7.2.10, we have that $\pi_k(S^{2n+1}) \cong \pi_k(\mathbb{C}P^n)$ for every $k \geq 3$. In particular, since $\mathbb{C}P^1 = S^2$, we have that $\pi_3(S^2) \cong \pi_3(S^3)$. ◇

We have the following general finiteness properties of homotopy groups

Theorem 7.2.12. *Let X be a connected finite CW-complex.*

(a) *If X is simply connected, then $\pi_n(X)$ is a finitely generated abelian group for each n.*
(b) *In general, $\pi_n(X, x_0)$ is finitely generated as a $\mathbb{Z}\pi_1(X, x_0)$ module.*

Example 7.2.13. Let X be the space

Then $\pi_2(X) \cong \pi_2(\tilde{X})$ where $\tilde{X}$, the universal cover of X, is

and $\pi_2(\tilde{X})$ is not finitely generated. ◇

Theorem 7.2.14. $\pi_i(S^n) = 0$ *for* $i < n$.

Proof. Give S^i a CW-structure with one cell in dimension i and one cell in dimension 0, and give S^n a CW-structure with one cell in dimension n and one cell in dimension 0. Let $f : S^i \to S^n$ represent an element of $\pi_i(S^n)$. Then by Theorem 4.2.28, f is (freely) homotopic to a cellular map $g : S^i \to S^n$. But by the definition of a cellular map, the image of the i-skeleton of S^i must be contained in the i-skeleton of S^n, which is a point. Thus f is freely homotopic to a constant map. But by Corollary 2.3.3 $\pi_1(S^n) = 0$ for $n > 1$, so by Corollary 7.1.12 f is homotopic as a map of pairs to a constant map. □

(This proof is deceptively simple, as the proof of Theorem 4.2.28 is highly nontrivial.)

Theorem 7.2.15 (Hopf). *Two maps $f : S^n \to S^n$ and $g : S^n \to S^n$ are homotopic if and only if they have the same degree.*

Corollary 7.2.16. *For any $n \geq 1$, $\pi_n(S^n) \cong \mathbb{Z}$.*

Proof. By Hopf's theorem, we have an isomorphism

$$\pi_n(S^n) \longrightarrow \{\text{degrees of maps from } S^n \text{ to } S^n\}.$$

But this latter set is $\mathbb{Z}$ by Theorem 4.2.31.   □

Hopf's theorem has a vast generalization due to Hurewicz, which we now give.

Definition 7.2.17. The *Hurewicz map* $\theta_n : \pi_n(X,x_0) \to H_n(X)$ or $\theta_n : \pi_n(X,A,x_0) \to H_n(X,A)$ is defined as follows: Let $f : (S^n,1) \to (X,x_0)$ represent $\alpha \in \pi_n(X,x_0)$. Let σ_n be the standard generator of $H_n(S^n)$ as defined in Remark 4.1.10. Then

$$\theta_n(\alpha) = f_*(\sigma_n).$$

Similarly, if $f : (D^n, S^{n-1}, 1) \to (X,A,x_0)$ represents $\alpha \in \pi_n(X,x_0)$, then $\theta_n(\alpha) = f_*(\delta_n)$.   ◇

Note that $\theta_1 : \pi_1(X,x_0) \to H_1(X)$ is the map θ of Sect. 5.2.

Lemma 7.2.18. *The following diagram commutes:*

$$
\begin{array}{ccccccccc}
\cdots \longrightarrow & \pi_n(A,x_0) & \longrightarrow & \pi_n(X,x_0) & \longrightarrow & \pi_n(X,A,x_0) & \overset{\partial}{\longrightarrow} & \pi_{n-1}(A,x_0) & \longrightarrow \cdots \\
& \downarrow{\scriptstyle \theta_n} & & \downarrow{\scriptstyle \theta_n} & & \downarrow{\scriptstyle \theta_n} & & \downarrow{\scriptstyle \theta_{n-1}} & \\
\cdots \longrightarrow & H_n(A) & \longrightarrow & H_n(X) & \longrightarrow & H_n(X,A) & \overset{\partial}{\longrightarrow} & H_{n-1}(A) & \longrightarrow \cdots
\end{array}
$$

Theorem 7.2.19 (Hurewicz). *Let X be a path-connected space. For any fixed integer $n \geq 2$ the following are equivalent:*

(a) $\pi_1(X) = 0$ *and* $H_k(X) = 0$ *for* $k = 1, \ldots, n-1$.
(b) $\pi_k(X) = 0$ *for* $k = 1, \ldots, n-1$.

In this situation, the Hurewicz map

$$\theta_n : \pi_n(X) \longrightarrow H_n(X)$$

is an isomorphism.

Definition 7.2.20. The space X is *n-connected* if $\pi_n(X,x_0) = 0$ for $i < n$. The pair (X,A) is *n-connected* if the map $\pi_0(A) \to \pi_0(X)$ is onto and $\pi_i(X,A) = 0$ for $1 \leq i \leq n$.   ◇

Recall we considered the suspension ΣX of a space X in connection with our discussion of homology. We now consider it in connection with homotopy.

Theorem 7.2.21. (a) *For any space X, ΣX is path-connected.*
(b) *For any path-connected space X, ΣX is simply connected.*
(c) *If X is n-connected for any $n \geq 1$, then ΣX is $(n+1)$-connected.*

Proof. (a) is trivial and (b) follows from van Kampen's theorem. (c) is a consequence of Theorem 3.2.13 and Hurewicz's theorem.   □

Now let $f : X \to Y$ be arbitrary. Then we have the map $\Sigma f : \Sigma X \to \Sigma Y$. In particular we may let $X = S^k$, so $f : S^k \to Y$ gives $\Sigma f : \Sigma S^k \to Y$, and we may naturally identify ΣS^k with S^{k+1}. Clearly if f and g are homotopic, so are Σf and Σg, and so we obtain a map $\Sigma : \pi_k(Y, y_0) \to \pi_{k+1}(\Sigma Y, y_0)$.

Theorem 7.2.22 (Freudenthal suspension theorem). *Let $n \geq 2$ and suppose that Y is $(n-1)$-connected. Then $\Sigma : \pi_k(Y) \to \pi_{k+1}(Y)$ is an isomorphism if $k \leq 2n-2$ and an epimorphism if $k = 2n-1$.*

Corollary 7.2.23. (a) *For any space Y, and any integer k, the sequence*

$$\pi_k(Y) \xrightarrow{\Sigma} \pi_{k+1}(Y) \xrightarrow{\Sigma} \pi_{k+2}(Y) \xrightarrow{\Sigma} \cdots$$

consists of isomorphisms from some point on.
(b) *In particular, taking $Y = S^n$, $\Sigma : \pi_{k+n}(S^n) \to \pi_{k+n+1}(S^{n+1})$ is an epimorphism for $n = k+1$ and an isomorphism for $n \geq k+2$.*

Definition 7.2.24. The limit group in the sequence in (b) is called the *stable k-stem*, denoted π_k^s.   ◊

We conclude by summarizing some basic facts about the homotopy groups of spheres, a field that has been an active area of research for almost 80 years, and is still going strong.

Theorem 7.2.25. (a) $\pi_1(S^1) \cong \mathbb{Z}$ *and* $\pi_n(S^1) = 0$ *for* $n > 1$.
(b) $\pi_i(S^n) = 0$ *for* $i < n$.
(c) $\pi_n(S^n) \cong \mathbb{Z}$.
(d) $\pi_i(S^3) \cong \pi_i(S^2)$ *for* $i \geq 3$.
(e) *(Serre) For n even, $\pi_{2n-1}(S^n) \cong \mathbb{Z} \oplus$ a finite group.*
(f) *(Serre) Except for cases (c) and (e), $\pi_i(S^n)$ is a finite group.*
(g) $\pi_1^s \cong \mathbb{Z}_2$, $\pi_2^s \cong \mathbb{Z}_2$, $\pi_3^s \cong \mathbb{Z}_{24}$.

7.3 Exercises

Exercise 7.3.1. Prove Theorem 7.1.9(4), the exactness of the homotopy sequence of a pair.

Exercise 7.3.2. Prove Lemma 7.1.17.

Exercise 7.3.3. Prove the claim of Example 7.2.7(3), that $p : S^{2n+1} \to \mathbb{C}P^n$ is a locally trivial fiber bundle with fiber S^1.

Exercise 7.3.4. Regard $\mathbb{C}P^n$ as a subspace of $\mathbb{C}P^{n+1}$ by identifying the point $[z_0, \ldots, z_n]$ of $\mathbb{C}P^n$ with the point $[z_0, \ldots, z_n, 0]$ of $\mathbb{C}P^{n+1}$. Let $\mathbb{C}P^\infty = \bigcup_{n=0}^\infty \mathbb{C}P^n$. Show that $\pi_k(\mathbb{C}P^\infty) = \mathbb{Z}$ for $k = 2$ and $\pi_k(\mathbb{C}P^\infty) = 0$ for $k \neq 2$.

Appendix A
Elementary Homological Algebra

Homological algebra is the branch of algebra that arose out of the necessity to provide algebraic foundations for homology theory. In this appendix we do not propose to systematically develop homological algebra. Rather, we just wish to develop it far enough to provide for our needs in this book. Thus, we do not advise the reader to read it straight through – it will seem like a curious collection of unmotivated results – but rather to refer to it as necessary.

The basic method of proof here is "diagram chasing". We do some representative proofs in detail but leave most of them for the reader.

Throughout this appendix R denotes an arbitrary commutative ring with 1. We will specialize to $R = \mathbb{Z}$, or a field, as necessary.

A.1 Modules and Exact Sequences

Definition A.1.1. A sequence of R-modules

$$\cdots \longrightarrow A_{i-1} \xrightarrow{\varphi_{i-1}} A_i \xrightarrow{\varphi_i} A_{i+1} \xrightarrow{\varphi_{i+1}} A_{i+2} \longrightarrow \cdots$$

is *exact* if for each i, $\mathrm{Ker}\,(\varphi_{i+1}) = \mathrm{Im}\,(\varphi_i)$. ◇

Remark A.1.2. In an exact sequence, $A_i = 0$ is equivalent to $\varphi_{i-2} : A_{i-2} \to A_{i-1}$ being a surjection and $\varphi_{i+1} : A_{i+1} \to A_{i+2}$ being an injection. ◇

Remark A.1.3. A sequence $0 \to A \to 0$ is exact if and only if $A = 0$. A sequence $0 \to A \xrightarrow{\varphi} B \to 0$ is exact if and only if φ is an isomorphism. ◇

Definition A.1.4. An exact sequence of R-modules

$$0 \longrightarrow A \xrightarrow{\varphi} B \xrightarrow{\psi} C \longrightarrow 0$$

is a *short exact sequence*. ◇

© Springer International Publishing Switzerland 2014
S.H. Weintraub, *Fundamentals of Algebraic Topology*, Graduate Texts in Mathematics 270, DOI 10.1007/978-1-4939-1844-7

Remark A.1.5. This is equivalent to: φ is an injection, ψ is a surjection, and $\mathrm{Im}\,(\varphi) = \mathrm{Ker}\,(\psi)$.   ◊

Definition A.1.6. A short exact sequence of R-modules

$$0 \longrightarrow A \xrightarrow{\varphi} B \xrightarrow{\psi} C \longrightarrow 0$$

is *split* if $\mathrm{Im}\,(\varphi) = \mathrm{Ker}\,(\psi)$ is a direct summand of B.   ◊

Lemma A.1.7. *Given a short exact sequence of R-modules*

$$0 \longrightarrow A \xrightarrow{\varphi} B \xrightarrow{\psi} C \longrightarrow 0$$

the following are equivalent:

(1) *There exists a homomorphism (necessarily a surjection) $\alpha : B \to A$ such that $\alpha \circ \varphi = id_A$.*
(2) *There exists a homomorphism (necessarily an injection) $\beta : C \to B$ such that $\psi \circ \beta = id_C$.*
(3) *This sequence is split.*

If these equivalent conditions hold then α and β are said to split (or be a splitting of) the sequence, and

$$B \cong \mathrm{Im}\,(\varphi) \oplus \mathrm{Ker}\,(\alpha)$$
$$\cong \mathrm{Ker}\,(\psi) \oplus \mathrm{Im}\,(\beta)$$
$$\cong A \oplus C.$$

Lemma A.1.8 (The five lemma). *Given a commutative diagram of exact sequences*

$$
\begin{array}{ccccccccc}
A_1 & \longrightarrow & A_2 & \longrightarrow & A_3 & \longrightarrow & A_4 & \longrightarrow & A_5 \\
\downarrow{\scriptstyle f_1} & & \downarrow{\scriptstyle f_2} & & \downarrow{\scriptstyle f_3} & & \downarrow{\scriptstyle f_4} & & \downarrow{\scriptstyle f_5} \\
B_1 & \longrightarrow & B_2 & \longrightarrow & B_3 & \longrightarrow & B_4 & \longrightarrow & B_5
\end{array}
$$

If f_1, f_2, f_4, and f_5 all isomorphisms, then so is f_3.

Proof. First we show f_3 is an injection. Let $x \in A_3$ with $f_3(x) = 0$, i.e., x "goes to 0" in B_3. Then x goes to 0 in B_4. Now x goes to some element y in A_4. By commutativity y goes to 0 in B_4. But f_4 is an isomorphism, so $y = 0$. Thus x goes to 0 in A_4, so by exactness x comes from some z in A_2. Then z goes to some w in B_2. By commutativity w goes to 0 in B_3, so w comes from some v in B_1. Since f_1 is an isomorphism v comes from some u in A_1. Then u goes to some t in A_2 and by commutativity t also goes to w in B_2. But f_2 is an isomorphism, so $t = z$. Thus u in A_1 goes to z in A_2 which goes to x in A_3. By the exactness of the top row, $x = 0$.

Next we show f_3 is a surjection. Let $x \in B_3$. Then x goes to some y in B_4. Since f_4 is an isomorphism, y comes from some z in A_4. By exactness, y goes to 0 in B_5, so z also goes to 0 in B_5. But f_5 is an isomorphism, so z goes to 0 in A_5. By exactness, z comes from some w in A_3. Then w goes to $v = f_3(w)$ in B_3. By commutativity, v also goes to y in B_4, so $x - v$ goes to 0 in B_4. Hence $x - v$ comes from some u in B_2. Since f_2 is an isomorphism, u comes from some t in A_2. Then t goes to some s in A_3, and by commutativity s goes to $x - v$ in B_3, i.e., $f(s) = x - v$. But then $f(s+w) = x - v + v = x$. □

Corollary A.1.9 (The short five lemma). *Given a commutative diagram of short exact sequences*

$$
\begin{array}{ccccccccc}
0 & \longrightarrow & A_2 & \longrightarrow & A_3 & \longrightarrow & A_4 & \longrightarrow & 0 \\
& & \downarrow{f_2} & & \downarrow{f_3} & & \downarrow{f_4} & & \\
0 & \longrightarrow & B_2 & \longrightarrow & B_3 & \longrightarrow & B_4 & \longrightarrow & 0
\end{array}
$$

If f_2 and f_4 are isomorphisms, then so is f_3.

Recall the following basic definition.

Definition A.1.10. An R-module M is *free* if there is some subset $S = \{m_i\}_{i \in I}$ of M such that every $m \in M$ can be expressed uniquely as a finite sum

$$
m = \sum_{i \in I} r_i m_i, \quad r_i \in R, \text{ only finitely many nonzero.}
$$

In this case, S is a *basis* of M. ◊

Lemma A.1.11. (a) *Let C be a free R-module. Then every short exact sequence of R-modules*

$$
0 \longrightarrow A \longrightarrow B \longrightarrow C \longrightarrow 0
$$

is split.
(b) *If $R = \mathbb{F}$ is a field, every short exact sequence of R-modules is split.*

Proof. (a) Let C have basis $\{c_i\}_{i \in I}$. For each i, let $b_i \in B$ with $\psi(b_i) = c_i$. Then there is a unique map $\beta : C \to B$ defined by $\beta(c_i) = b_i$ for each i, and this gives a splitting by Lemma A.1.7.
(b) If $R = \mathbb{F}$ is a field, an R-module is an $\mathbb{F}$-vector space, so has a basis, and so is free.

□

We remind the reader of the construction of the dual of a module and the dual of a map, which are at the core of cohomology.

Definition A.1.12. Let M be an R-module. Its *dual module* M^* is the R-module

$$M^* = \operatorname{Hom}(M, R),$$

the module of R-homomorphisms from M to R.

If $f : M \to N$ is a map of R-modules, the *dual map* $f^* : N^* \to M^*$ is the map defined by

$$(f^*(\alpha))(m) = \alpha(f(m)) \quad \text{for } \alpha \in N^*, \; m \in M. \qquad \Diamond$$

We now generalize the notions of ring and module to the graded case.

Definition A.1.13. (1) A *graded commutative ring* $\mathscr{S}$ is a ring $\mathscr{S}$ with 1 such that the additive group of $\mathscr{S}$ decomposes as $\mathscr{S} = \bigoplus_{i \in \mathbb{Z}} S_i$ and the multiplication has the property that of $x \in S_j$ and $y \in S_k$ then $xy \in S_{j+k}$. Also, the ordinary commutative law for multiplication is replaced by the law

$$yx = (-1)^{jk}xy \quad \text{for } x \in S_j, \; y \in S_k.$$

(2) A *left module* $\mathscr{N}$ over a graded commutative ring $\mathscr{S}$ is a left $\mathscr{S}$-module $\mathscr{N}$ such that $\mathscr{N} = \bigoplus_{i \in \mathbb{Z}} N_i$ and the module structure has the property that if $s \in S_j$ and $n \in N_{j+k}$ then $sn \in N_k$.

(3) A *ring homomorphism* $\varphi : \mathscr{S} \to \mathscr{T}$, where $\mathscr{N} = \bigoplus_{i \in \mathbb{Z}} S_i$ and $\mathscr{T} = \bigoplus_{i \in \mathbb{Z}} T_i$ are graded commutative rings, is a homomorphism of rings that satisfies the additional property that if $s_i \in S_i$, then $t_i = \varphi(s_i) \in T_i$.

(4) A *graded commutative R-algebra* $\mathscr{S}$ is a graded commutative ring that is an R-algebra with the property that if $r \in R$ and $s_i \in S_i$, then $rs_i \in S_i$.

(5) An *algebra homomorphism* $\varphi : \mathscr{S} \to \mathscr{T}$ between graded commutative R-algebras is a ring homomorphism that is a map of algebras. $\qquad \Diamond$

(Of course, any graded commutative ring is a graded commutative $\mathbb{Z}$-algebra.)

A.2   Chain Complexes

Definition A.2.1. A *chain complex* over R is $\mathscr{A} = \{A_i, d_i\}_{i \in \mathbb{Z}}$ a set of R-modules A_i, and R-module homomorphisms $d_i : A_i \to A_{i-1}$, with the property that $d_{i-1}d_i : A_i \to A_{i-2}$ is the 0 map for every i. $\qquad \Diamond$

Often we abbreviate d_i by d, and write the relation $d_{i-1}d_i = 0$ as $d^2 = 0$. Note that this condition implies $\operatorname{Im}(d_{i+1}) \subseteq \operatorname{Ker}(d_i)$ for every i.

Definition A.2.2. Let $\mathscr{A}$ be a chain complex. Then the i-th *homology group* $H_i(\mathscr{A})$ is the R-module

$$H_i(\mathscr{A}) = \operatorname{Ker}(d_i)/\operatorname{Im}(d_{i+1}). \qquad \Diamond$$

Definition A.2.3. Let $a \in \mathrm{Ker}\,(d_i)$. Then $[a]$ denotes the image of a in $H_i(\mathscr{A})$ under the quotient map $\mathrm{Ker}\,(d_i) \to \mathrm{Ker}\,(d_i)/\mathrm{Im}\,(d_{i+1})$, and $[a]$ is the *homology class represented by a*. If $\alpha \in H_i(\mathscr{A})$ and $a \in \mathrm{Ker}\,(d_i)$ with $[a] = \alpha$, then a *represents* (or is a *representative* of) α. ◊

Definition A.2.4. An element $a \in A_i$ is called a *chain*. An element $a \in \mathrm{Ker}\,(d_i)$ is called a *cycle*. An element $a \in \mathrm{Im}\,(d_{i+1})$ is called a *boundary*. ◊

Definition A.2.5. Let $\mathscr{A} = \{A_i, d_i^A\}$ and $\mathscr{B} = \{B_i, d_i^B\}$ be chain complexes. A *map of chain complexes* $F : \mathscr{A} \to \mathscr{B}$ is a collection of homomorphisms $F = \{f_i : A_i \to B_i\}_{i \in \mathbb{Z}}$ such that for each i, the following diagram commutes:

$$
\begin{array}{ccc}
A_i & \xrightarrow{\ f_i\ } & B_i \\
{\scriptstyle d_i^A}\downarrow & & \downarrow{\scriptstyle d_i^B} \\
A_{i-1} & \xrightarrow{\ f_{i-1}\ } & B_{i-1}.
\end{array}
$$

◊

Definition A.2.6. Let $F : \mathscr{A} \to \mathscr{B}$ be a map of chain complexes. The *induced map on homology* $F_* : H_*(\mathscr{A}) \to H_*(\mathscr{B})$ is defined as follows: $F_* = \{f_i : H_i(\mathscr{A}) \to H_i(\mathscr{B})\}_{i \in \mathbb{Z}}$ where $f_i([a]) = [f_i(a)]$. ◊

Lemma A.2.7. *The induced map on homology F_* is well-defined.*

Proof. We must show it is independent of the choice of representative a. Thus suppose $[a] = [a']$. Then $a = a' + a''$ where $a'' \in \mathrm{Im}\,(d_{i+1})$, i.e., $a'' = d_{i+1}(a''')$ for some a'''. But then

$$
\begin{aligned}
f_i(a) = f_i(a' + a'') &= f_i(a') + f_i(a'') \\
&= f_i(a') + f_i(d_{i+1}(a''')) \\
&= f_i(a') + d_{i+1}(f_i(a'''))
\end{aligned}
$$

so $[f_i(a)] = [f_i(a')]$. □

Definition A.2.8. Let $F : \mathscr{A} \to \mathscr{B}$ and $G : \mathscr{A} \to \mathscr{B}$ be maps of chain complexes. A *chain homotopy* between F and G is a collection of maps $\Phi = \{\varphi_i : A_i \to B_{i+1}\}_{i \in \mathbb{Z}}$ such that

$$
d_{i+1}^B \varphi_i + \varphi_{i-1} d_i^A = f_i - g_i : A_i \longrightarrow B_i \quad \text{for each } i
$$

◊

Lemma A.2.9. *Suppose that there is a chain homotopy Φ between F and G. Then $F_* = G_*$, i.e., $f_i = g_i$ for each i.*

Proof. Let $a \in A_i$ be a cycle. Then

$$
[f_i(a)] = [(g_i + d_{i+1}^B \varphi_i + \varphi_{i-1} d_i^A)(a)]
$$

$$= [g_i(a)] + [d^B_{i+1}\varphi_i(a)] + [\varphi_{i-1}d^A_i(a)]$$
$$= [g_i(a)]$$

as $d^B_{i+1}(\varphi_i(a))$ is a boundary and $d^A_i(a) = 0$ since a is a cycle. $\qquad\square$

Theorem A.2.10. *Let* $0 \to \mathscr{A} \xrightarrow{F} \mathscr{B} \xrightarrow{G} \mathscr{C} \to 0$ *be a short exact sequence of chain complexes, i.e., suppose that for each i,* $0 \to A_i \xrightarrow{f_i} B_i \xrightarrow{g_i} C_i \to 0$ *is exact. Then there is a long exact sequence in homology*

$$\cdots \longrightarrow H_i(\mathscr{A}) \xrightarrow{f_i} H_i(\mathscr{B}) \xrightarrow{g_i} H_i(\mathscr{C}) \xrightarrow{\partial_i} H_{i-1}(\mathscr{A}) \longrightarrow \cdots .$$

Proof. We show how to define ∂ by a "diagram chase". The remainder of the proof is a further diagram chase, which we omit.

We have:

$$
\begin{array}{ccccccccc}
0 & \longrightarrow & A_i & \xrightarrow{f_i} & B_i & \xrightarrow{g_i} & C_i & \longrightarrow & 0 \\
& & \downarrow{d^A_i} & & \downarrow{d^B_i} & & \downarrow{d^C_i} & & \\
0 & \longrightarrow & A_{i-1} & \xrightarrow{f_{i-1}} & B_{i-1} & \xrightarrow{g_{i-1}} & C_{i-1} & \longrightarrow & 0
\end{array}
$$

Let $c_i \in C_i$ be a cycle. Since g_i is onto, there is an element $b_i \in B_i$ with $g_i(b_i) = c_i$. Let $b_{i-1} = d^B_i(b_i)$. Then $g_{i-1}(b_{i-1}) = g_{i-1}d^B_i(b_i) = d^C_i g_i(b_i)$ by commutativity $= d^C_i(c_i) = 0$ since c_i is a cycle. By exactness, there is a unique $a_{i-1} \in A_{i-1}$ with $f_{i-1}(a_{i-1}) = b_i$. Define

$$\partial_i([c_i]) = [a_{i-1}] \in H_{i-1}(\mathscr{A}).$$

$\qquad\square$

Theorem A.2.11. *Suppose there is a commutative diagram of exact sequences*

$$
\begin{array}{ccccccccc}
\cdots & \longrightarrow & A_i & \xrightarrow{\alpha_1} & B_i & \xrightarrow{\gamma_1} & C_i & \longrightarrow & A_{i-1} & \longrightarrow & \cdots \\
& & \downarrow{\alpha_2} & & \downarrow{\beta_1} & & \downarrow{\varepsilon} & & \downarrow{\alpha_2} & & \\
\cdots & \longrightarrow & D_i & \xrightarrow{\beta_2} & E_i & \xrightarrow{\gamma_2} & F_i & \longrightarrow & D_{i-1} & \longrightarrow & \cdots
\end{array}
$$

and suppose further that ε *is an isomorphism, for each i.*
    Then there is a long exact sequence

$$\cdots \longrightarrow A_i \xrightarrow{\alpha} B_i \oplus D_i \xrightarrow{\beta} E_i \xrightarrow{\Delta} A_{i-1} \longrightarrow \cdots$$

where the maps α, β, Δ are defined by:

$$\alpha(q) = (\alpha_1(q), \alpha_2(q))$$
$$\beta(r,s) = \beta_1(r) - \beta_2(s)$$
$$\Delta(t) = \partial_1 \varepsilon^{-1} \gamma_2(t).$$

Proof. We shall chase this diagram to show that the given sequence is exact.

(i) $\mathrm{Im}(\alpha) \subseteq \mathrm{Ker}(\beta)$: Let $(r,s) = \alpha(q) = (\alpha_1(q), \alpha_2(q))$ for some q. Then $\beta(r,s) = \beta_1(r) - \beta_2(s) = \beta_1\alpha_1(q) - \beta_2\alpha_2(q) = 0$ by the commutativity of the diagram.

(ii) $\mathrm{Im}(\beta) \subseteq \mathrm{Ker}(\Delta)$: Let $t = \beta(r,s) = \beta_1(r) - \beta_2(s)$ for some r,s. Then $\Delta(t) = \partial_1\varepsilon^{-1}\gamma_2(t) = \partial_1\varepsilon^{-1}\gamma_2(\beta_1(r)) - \partial_1\varepsilon^{-1}\gamma_2(\beta_2(s))$. But, by commutativity, $\gamma_1 = \varepsilon^{-1}\gamma_2\beta_1$ so $\Delta(t) = \partial_1\gamma_1(r) - \partial_1\varepsilon^{-1}(\gamma_2\beta_2(s)) = 0 - 0 = 0$ by exactness.

(iii) $\mathrm{Im}(\Delta) \subseteq \mathrm{Ker}(\alpha)$: Let $q = \partial_1\varepsilon^{-1}\gamma_2(t)$. Then $\alpha(q) = (\alpha_1\partial_1\varepsilon^{-1}\gamma_2(t), \alpha_2\partial_1\varepsilon^{-1}\gamma_2(t))$. But, by commutativity, $\partial_2 = \alpha_2\partial_1\varepsilon^{-1}$ so $\alpha(q) = (\alpha_1\partial_1(\varepsilon^{-1}\gamma_2(t)), \partial_2\gamma_2(t)) = (0,0) = 0$.

(iv) $\mathrm{Ker}(\beta) \subseteq \mathrm{Im}(\alpha)$: Suppose $\beta(r,s) = \beta_1(r) - \beta_2(s) = 0$. Let $u = \beta_1(r) = \beta_2(s)$. Then $\gamma_2(u) = \gamma_2\beta_1(r) = \gamma_2\beta_2(s) = 0$ by exactness. By commutativity, $\gamma_2\beta_1(r) = \varepsilon\gamma_1(r)$ so $\varepsilon\gamma_1(r) = 0$. But ε is an isomorphism, so $\gamma_1(r) = 0$. Then, by exactness, $r = \alpha_1(q_0)$ for some q_0. Let $s_0 = \alpha_2(q_0)$. Then $\beta_2(s_0) = \beta_2\alpha_2(q_0) = \beta_1\alpha_1(q_0) = \beta_1(r) = \beta_2(s)$, so $\beta_2(s - s_0) = 0$. Then, by exactness, $s - s_0 = \partial_2(v)$ for some v. Let $w = \varepsilon^{-1}(v)$ and $x = \partial_1(w)$. Then $\alpha_2(x) = \alpha_2\partial_1(w) = \partial_2\varepsilon(w) = \partial_2\varepsilon(\varepsilon^{-1}(v)) = \partial_2(v) = s - s_0$. Set $q = q_0 + x$. Then $\alpha_1(q) = \alpha_1(q_0 + x) = \alpha_1(q_0) + \alpha_1(x) = \alpha_1(q_0) + \alpha_1\partial_1(w) = \alpha_1(q_0) = r$, and $\alpha_2(q) = \alpha_2(q_0 + x) = \alpha_2(q_0) + \alpha_2(x) = s_0 + (s - s_0) = s$. Thus $(r,s) = \alpha(q)$.

(v) $\mathrm{Ker}(\Delta) \subseteq \mathrm{Im}(\beta)$: Suppose $\Delta(t) = \partial_1\varepsilon^{-1}\gamma_2(t) = 0$. Then $0 = \partial_1(\varepsilon^{-1}\gamma_2(t))$ so $\varepsilon^{-1}\gamma_2(t) = \gamma_1(r)$ for some r. Then $\varepsilon\gamma_1(r) = \gamma_2(t)$, and then $\gamma_2\beta_1(r) = \varepsilon\gamma_1(r) = \gamma_2(t)$. Hence $\gamma_2(\beta_1(r) - t) = 0$ so $\beta_1(r) - t = \beta_2(s)$ for some s. But then $\beta_1(r) - \beta_2(s) = t$, so $t = \beta(r,s)$.

(vi) $\mathrm{Ker}(\alpha) \subseteq \mathrm{Im}(\Delta)$: Suppose $\alpha(q) = 0$. Then $\alpha_1(q) = \alpha_2(q) = 0$. Since $\alpha_1(q) = 0$, $q = \partial_1(p)$ for some p. Set $n = \varepsilon(p)$. Then $\partial_2(n) = \partial_2\varepsilon(p) = \alpha_2\partial_1(p) = \alpha_2(q) = 0$, so $n = \gamma_2(t)$ for some t. But then $\Delta(t) = \partial_1\varepsilon^{-1}\gamma_2(t) = \partial_1\varepsilon^{-1}(n) = \partial_1(p) = q$.

□

Theorem A.2.12. *Given a commutative diagram of exact sequences*

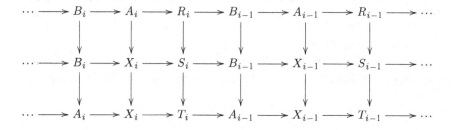

where the vertical maps $B_i \to B_i$ and $X_i \to X_i$ are both identity maps, the vertical map $B_i \to A_i$ agrees with the horizontal map $B_i \to A_i$, and the vertical map $A_i \to X_i$ agrees with the horizontal map $A_i \to X_i$, for all i, there is a long exact sequence

$$\cdots \longrightarrow R_i \longrightarrow S_i \longrightarrow T_i \overset{\partial}{\longrightarrow} A_{i-1} \longrightarrow \cdots$$

where the map ∂ is the composition of $T_i \to A_{i-1}$ on the bottom row, followed by the identity map from A_{i-1} on the bottom row to A_{i-1} on the top row, followed by $A_{i-1} \to R_{i-1}$ on the top row.

Proof. We shall observe that under the hypotheses of the theorem, the diagram continues to commute if we insert the diagonal arrows from A_i on the bottom row to A_i on the top row, with these maps being the identity maps.

Otherwise, the proof is a very elaborate diagram chase, which we omit. □

A finitely generated abelian group A is isomorphic to $F \oplus T$ where F is a free abelian group of well-defined rank r (i.e., F is isomorphic to $\mathbb{Z}^r$) and T is a torsion group. In this case we define the rank of A to be r.

Theorem A.2.13. *Let $0 \to C_n \overset{\partial_n}{\to} C_{n-1} \overset{\partial_{n-1}}{\longrightarrow} \cdots \to C_1 \overset{\partial_1}{\to} C_0 \to 0$ be a chain complex with each C_i a finitely generated free abelian group of rank d_i. Let H_i be the i-th homology group of this chain complex, $i = 0, \ldots, n$. Then*

$$\sum_{i=0}^{n} (-1)^i \operatorname{rank} H_i = \sum_{i=0}^{n} (-1)^i \operatorname{rank} C_i.$$

Proof. Although, with a little care, this theorem can be proven directly, it is easiest to tensor everything with $\mathbb{Q}$. We then obtain a chain complex

$$0 \longrightarrow V_n \longrightarrow V_{n-1} \longrightarrow \cdots \longrightarrow V_1 \longrightarrow V_0 \longrightarrow 0$$

with $V_i = C_i \otimes \mathbb{Q}$ a rational vector space of dimension d_i. It is easy to check that if $\{K_i\}$ are the homology groups of the new chain complex, then $K_i = H_i \otimes \mathbb{Q}$ for each i. In particular, if $r_i = \operatorname{rank} H_i$, then K_i is a rational vector space of dimension r_i. Hence it suffices to prove that

$$\sum_{i=0}^{n} (-1)^{r_i} = \sum_{i=0}^{n} (-1)^{d_i}.$$

We prove this by induction on n. If $n = 0$, this is trivial. The chain complex is then $0 \to V_0 \to 0$ which has the single nonzero homology group $K_0 = V_0$, so certainly $r_0 = d_0$.

Suppose it is true for $n - 1$, and all chain complexes.

We have $\partial_n : V_n \to \operatorname{Im}(V_n) = V'_{n-1} \subseteq \operatorname{Ker}(\partial_{n-1}) \subseteq V_{n-1}$. Let V''_{n-1} be a complement of V'_{n-1} in $\operatorname{Ker}(\partial_{n-1})$ and let V'''_{n-1} be a complement of $\operatorname{Ker}(\partial_{n-1})$ in V_{n-1}. Then

$$V_{n-1} = V'_{n-1} \oplus V''_{n-1} \oplus V'''_{n-1}.$$

Let these three subspaces have dimensions $d'_{n-1}, d''_{n-1}, d'''_{n-1}$, respectively.

Then $\partial_{n-1}|V'''_{n-1}$ is an injection and $\partial_{n-1}(V'''_{n-1}) = \partial_{n-1}(V_{n-1})$. Thus we have a chain complex

$$0 \longrightarrow V'''_{n-1} \longrightarrow V_{n-2} \longrightarrow \cdots \longrightarrow V_1 \longrightarrow V_0 \longrightarrow 0$$

whose homology in dimension $n - 1$ is 0, and whose homology in dimension $i < n - 1$ is K_i. By the $n - 1$ case, we have

$$\sum_{i=0}^{n-2}(-1)^i r_i = \left(\sum_{i=0}^{n-2}(-1)^i d_i\right) + (-1)^{n-1}\left(d'''_{n-1}\right).$$

Now $K_{n-1} = \mathrm{Ker}\,(\partial_{n-1})/\mathrm{Im}\,(\partial_n)$ is isomorphic to V''_{n-1}, so $r_{n-1} = d''_{n-1}$. Hence

$$\sum_{i=0}^{n-1}(-1)^i r_i = \left(\sum_{i=0}^{n-2}(-1)^i d_i\right) + (-1)^{n-1}\left(d''_{n-1} + d'''_{n-1}\right).$$

Now $K_n = \mathrm{Ker}\,(\partial_n)$. Since $\mathrm{Im}\,(\partial_n)$ has dimension d'_{n-1}, $\mathrm{Ker}\,(\partial_n)$ has dimension $d_n - d'_{n-1}$, i.e., $r_n = d_n - d'_{n-1}$. Hence

$$
\begin{aligned}
\sum_{i=0}^{n}(-1)^i r_i &= \left(\sum_{i=0}^{n-2}(-1)^i d_i\right) + (-1)^{n-1}\left(d''_{n-1} + d'''_{n-1}\right) + (-1)^n\left(d_n - d'_{n-1}\right) \\
&= \left(\sum_{i=0}^{n-2}(-1)^i d_i\right) + (-1)^{n-1}\left(d'_{n-1} + d''_{n-1} + d'''_{n-1}\right) + (-1)^n d_n \\
&= \sum_{i=0}^{n}(-1)^i d_i
\end{aligned}
$$

as claimed.

Thus by induction we are done.   □

A.3   Tensor Product, Hom, Tor, and Ext

In this section we recapitulate the basic properties of the tensor product and Hom, and introduce Tor and Ext. These constructions depend on the ring R, but we assume R is fixed, and in the interests of simplicity suppress R from our notation. *Except in Lemmas A.3.1 and A.3.2, we assume that R is* PID.

We begin with the easiest situation.

Lemma A.3.1. *Let* $0 \to A \to B \to C \to 0$ *be a split short exact sequence of R-modules. Then for any R-module M,*

(a)   *The sequence* $0 \to A \otimes M \to B \otimes M \to C \otimes M \to 0$ *is exact, and*
(b)   *The sequence* $0 \to \operatorname{Hom}(C,M) \to \operatorname{Hom}(B,M) \to \operatorname{Hom}(A,M) \to 0$ *is exact.*

Proof. This follows from Lemma A.1.7 and the facts that $(A \oplus C) \otimes M \cong (A \otimes M) \oplus (C \otimes M)$ and $\operatorname{Hom}(A \oplus C, M) \cong \operatorname{Hom}(A,M) \oplus \operatorname{Hom}(C,M)$.   □

Lemma A.3.2. *Let* $0 \to A \to B \to C \to 0$ *be a short exact sequence of R-modules. Then for any R-module M*

(a)   *The sequence* $A \otimes M \to B \otimes M \to C \otimes M \to 0$ *is exact*
(b)   *The sequence* $0 \to \operatorname{Hom}(C,M) \to \operatorname{Hom}(B,M) \to \operatorname{Hom}(A,M)$ *is exact.*

Note the difference between the split and the nonsplit case: In the nonsplit case we do not in general obtain five-term exact sequences. We now introduce Tor and *Ext*, which measure inexactness. But first we must introduce free resolutions.

Lemma A.3.3. *Let M be an R-module. Then there is a short exact sequence*

$$0 \longrightarrow F_1 \xrightarrow{\varphi} F_0 \xrightarrow{\psi} M \longrightarrow 0$$

where F_0 and F_1 are free R-modules.

Proof. We can certainly find a free module F_0 and an epimorphism $\psi : F_0 \to M$, as follows: Let $\{m_i\}_{i \in I}$ generate M. (We can certainly find a set of generators. For example, we could choose this set to be all the elements of M.) Let F_0 be the free R-module with basis $\{f_i\}_{i \in I}$ and let $\psi : F_0 \to M$ be defined by $\psi(f_i) = m_i$ for each $i \in I$.

Let $F_1 = \operatorname{Ker}(\psi)$ and let $\varphi : F_1 \to F_0$ be the inclusion. Then we certainly have a short exact sequence

$$0 \longrightarrow F_1 \xrightarrow{\varphi} F_0 \xrightarrow{\psi} M \longrightarrow 0.$$

Furthermore, *since R is a* PID, every submodule of a free module is free, so F_1 is free as well.   □

Remark A.3.4. Note we have crucially used the hypothesis that R is a PID. (It would take us far afield to consider what happens when R is not.)   ◊

Definition A.3.5. A sequence of R-modules as in Lemma A.3.3 is a *free resolution* of M.   ◊

Lemma A.3.6. *Let M and N be R-modules. Then there is a well-defined R-module T (i.e., independent of the choice of F_1 and F_0) such that, if $0 \to F_1 \to F_0 \to M \to 0$ is a free resolution of M, the sequence*

$$0 \longrightarrow T \longrightarrow F_1 \otimes N \longrightarrow F_0 \otimes N \longrightarrow M \otimes N \longrightarrow 0$$

is exact.

Definition A.3.7. The module T of Lemma A.3.6 is the *torsion product* of M and N,

$$T = \text{Tor}(M,N).$$

◊

We summarize the properties of the torsion product.

Lemma A.3.8. (a) *If M is free, $\text{Tor}(M,N) = 0$ for every N.*
(b) $\text{Tor}(M_1 \oplus M_2, N) \cong \text{Tor}(M_1, N) \oplus \text{Tor}(M_2, N)$.
(c) $\text{Tor}(M, N_1 \oplus N_2) \cong \text{Tor}(M, N_1) \oplus \text{Tor}(M, N_2)$.
(d) *If N is free, $\text{Tor}(M,N) = 0$ for every M.*
(e) *Let M be a cyclic R-module, $M \cong R/I_M$, I_M an ideal of R, and let N be a cyclic R-module, $N \cong R/I_N$, I_N an ideal of R. Let $I_M = (r_M)$, i.e., let I_M be the principal ideal consisting of the multiples of the element r_M of R and similarly let $I_N = (r_N)$. If $r_M = 0$ or $r_N = 0$, then $\text{Tor}(M,N) = 0$. Otherwise, $\text{Tor}(M,N) \cong R/I_T$ where $I_T = (r_T)$ and $r_T = \gcd(r_M, r_N)$.*
(f) *If M is torsion-free, $\text{Tor}(M,N) = 0$ for every N.*
(g) *If N is torsion-free, $\text{Tor}(M,N) = 0$ for every M.*
(h) $\text{Tor}(M,N) \cong \text{Tor}(N,M)$.
(i) *If $R = \mathbb{F}$ is a field, $\text{Tor}(M,N) = 0$ for every M,N.*

Lemma A.3.9. *Let $0 \to A \to B \to C \to 0$ be a short exact sequence of R-modules. Then for any R-module N, the sequence*

$$0 \to \text{Tor}(A,N) \to \text{Tor}(B,N) \to \text{Tor}(C,N) \to A \otimes N \to B \otimes N \to C \otimes N \to 0$$

is exact.

Lemma A.3.10. *Let M and N be R-modules. Then there is a well-defined R-module E (i.e., independent of the choice of F_1 and F_0) such that, if $0 \to F_1 \to F_0 \to M \to 0$ is a free resolution of M, the sequence*

$$0 \longrightarrow \text{Hom}(M,N) \longrightarrow \text{Hom}(F_0,N) \longrightarrow \text{Hom}(F_1,N) \longrightarrow E \longrightarrow 0$$

is exact.

Definition A.3.11. The module E of Lemma A.3.10 is the *extension product* of M and N,

$$E = \text{Ext}(M,N).$$

◊

We summarize the properties of the extension product.

Lemma A.3.12. (a) *If M is free, $\text{Ext}(M,N) = 0$ for every N.*
(b) $\text{Ext}(M_1 \oplus M_2, N) \cong \text{Ext}(M_1, N) \oplus \text{Ext}(M_2, N)$.
(c) $\text{Ext}(M, N_1 \oplus N_2) \cong \text{Ext}(M, N_1) \oplus \text{Ext}(M, N_2)$.
(d) *Let M be a cyclic R-module, $R \cong R/I_M$ with $I_M = (r_M)$. Then for any R-module N,*

$$\text{Ext}(M,N) \cong N/r_M N$$

where for $r \in R$, $rN = \{rn \mid n \in N\} \subseteq N$. In particular,

$$\text{Ext}(M,R) \cong R/r_M R \cong M.$$

If N is a cyclic R-module, $N \cong R/I_N$ with $I_N = (r_N)$, then

$$\text{Ext}(M,N) \cong R/r_E R$$

where $r_E = \gcd(r_M, r_N)$. As a special case

$$\text{Ext}(M,N) = 0$$

if r_M and r_N are relatively prime.

Lemma A.3.13. *Let $0 \to A \to B \to C \to 0$ be a short exact sequence of R-modules. Then for any R-module N, the sequences*

$$0 \longrightarrow \text{Hom}(C,N) \longrightarrow \text{Hom}(B,N) \longrightarrow \text{Hom}(A,N)$$
$$\longrightarrow \text{Ext}(C,N) \longrightarrow \text{Ext}(B,N) \longrightarrow \text{Ext}(A,N) \longrightarrow 0$$

and

$$0 \longrightarrow \text{Hom}(N,A) \longrightarrow \text{Hom}(N,B) \longrightarrow \text{Hom}(N,C)$$
$$\longrightarrow \text{Ext}(N,A) \longrightarrow \text{Ext}(N,B) \longrightarrow \text{Ext}(N,C) \longrightarrow 0$$

are both exact.

(We need almost all of the results of this section to compute Tor and Ext, which enter into computations of homology and cohomology. We do not need Lemmas A.3.9 and A.3.13, but have given them to complete the picture.)

Remark A.3.14. Note that if $R = \mathbb{F}$ is a field, this section is entirely superfluous. For in this case every short exact sequence of R-modules is split (Lemma A.1.11), and so $\text{Tor}(M,N) = 0$ and $\text{Ext}(M,N) = 0$ for any two R-modules M and N. $\Diamond$

Appendix B
Bilinear Forms

In this appendix we introduce bilinear forms and we state some of the basic classification theorems.

B.1 Definitions

Definition B.1.1. Let R be a commutative ring and let M be a free R-module. A *bilinear form*

$$\langle \, , \, \rangle : M \times M \longrightarrow R$$

is a function that is linear in both arguments, i.e., with the property that

(1) $\langle r_1 m_1 + r_2 m_2, m \rangle = r_1 \langle m_1, m \rangle + r_2 \langle m_2, m \rangle$
(2) $\langle m, r_1 m_1 + r_2 m_2 \rangle = r_1 \langle m, m_1 \rangle + r_2 \langle m, m_2 \rangle$

for all $r_1, r_2 \in R$ and all $m_1, m_2, m \in M$. $\diamond$

Note that bilinearity is precisely the condition we need to obtain

$$\langle \, , \, \rangle : M \otimes M \longrightarrow R.$$

The appropriate equivalence relation on bilinear forms is that of isometry.

Definition B.1.2. Let $\langle \, , \, \rangle$ be a bilinear form on M and $\langle\langle \, , \, \rangle\rangle$ be a bilinear form on N. An *isometry* between these two forms is an isomorphism $\varphi : M \to N$ such that

$$\langle\langle \varphi(m_1), \varphi(m_2) \rangle\rangle = \langle m_1, m_2 \rangle \quad \text{for all } m_1, m_2 \in M.$$

In this situation, the two forms are said to be *isometric*. $\diamond$

© Springer International Publishing Switzerland 2014
S.H. Weintraub, *Fundamentals of Algebraic Topology*, Graduate Texts
in Mathematics 270, DOI 10.1007/978-1-4939-1844-7

Definition B.1.3. Let $\langle\,,\,\rangle : M \otimes M \to R$ be a bilinear form. Then $\langle\,,\,\rangle$ defines a map $\alpha : M \to M^* = \operatorname{Hom}(M,R)$ by

$$\alpha(m_1)(m_2) = \langle m_1, m_2 \rangle \quad \text{for all } m_2 \in M$$

and a map $\beta : M \to M^*$ by

$$\beta(m_2)(m_1) = \langle m_1, m_2 \rangle \quad \text{for all } m_1 \in M.$$

The form $\langle\,,\,\rangle$ is *nonsingular* if α and β are isomorphisms. $\Diamond$

Definition B.1.4. The bilinear form $\langle\,,\,\rangle$ is *symmetric* if

$$\langle m_1, m_2 \rangle = \langle m_2, m_1 \rangle$$

for all $m_1, m_2 \in M$, and is *skew-symmetric* if

$$\langle m_1, m_2 \rangle = -\langle m_2, m_1 \rangle$$

for all $m_1, m_2 \in M$. $\Diamond$

Example B.1.5. Let $M = R^n$ and let A be any $n \times n$ matrix with entries in R. Then $[A]$ is the bilinear form

$$[A] = \langle\,,\,\rangle : M \times M \longrightarrow R \quad \text{by } \langle x, y \rangle = x^t A y$$

is a bilinear form. It is symmetric if $A = A^t$ and skew-symmetric if $A = -A^t$. It is nonsingular if A is nonsingular (i.e., invertible). $\Diamond$

Remark B.1.6. In the situation of Example B.1.5, either α an isomorphism or β an isomorphism implies A nonsingular, so in the finite rank case it is only necessary to check one of these conditions. $\Diamond$

Remark B.1.7. Upon choosing a basis of M, a module of finite rank, every bilinear form arises in this way. $\Diamond$

Here is a simple but basic construction.

Definition B.1.8. Let $\langle\,,\,\rangle$ be a bilinear form on M and let $\langle\,,\,\rangle'$ be a bilinear form on M'. Their direct sum $\langle\,,\,\rangle'' = \langle\,,\,\rangle \oplus \langle\,,\,\rangle'$ is the bilinear form on $M'' = M \oplus M'$ given by

$$\langle m_1 + m_1', m_2 + m_2' \rangle'' = \langle m_1, m_2 \rangle + \langle m_1', m_2' \rangle'$$

for all $m_1, m_2 \in M$ and $m_1', m_2' \in M'$. Also, $k\langle\,,\,\rangle$ denotes $\langle\,,\,\rangle \oplus \cdots \oplus \langle\,,\,\rangle$, where there are k summands. $\Diamond$

Remark B.1.9. If, in the notation of Example B.1.5, $\langle\,,\,\rangle = [A]$ and $\langle\,,\,\rangle' = [A']$, then $\langle\,,\,\rangle'' = \langle\,,\,\rangle \oplus \langle\,,\,\rangle' = [A'']$ with A'' the block diagonal matrix $A'' = \begin{bmatrix} A & 0 \\ 0 & A' \end{bmatrix}$. $\Diamond$

B.2   Classification Theorems

We now give some of the basic classification theorems for nonsingular bilinear forms. First we consider skew-symmetric forms.

Theorem B.2.1. *Let* $R = \mathbb{Z}$ *or a field of characteristic not equal to two. Then any nonsingular skew-symmetric bilinear form on a free R-module M of finite rank is isometric to*

$$k \begin{bmatrix} 0 & 1 \\ -1 & 0 \end{bmatrix}$$

for some integer k. In particular, if M admits a nonsingular skew-symmetric bilinear form, then rank(M) *is even.*

Next we consider symmetric forms over the real numbers $\mathbb{R}$.

Definition B.2.2. A symmetric bilinear form $\langle \, , \, \rangle$ on an $\mathbb{R}$-vector space V is positive definite if $\langle v, v \rangle > 0$ for every $v \in V$, $v \neq 0$, and is negative definite if $\langle v, v \rangle < 0$ for every $v \in V$, $v \neq 0$.   ◇

Lemma B.2.3. *Let* $\langle \, , \, \rangle$ *be a nonsingular symmetric bilinear form on a real vector space V of dimension n. Let* V_+ *be a subspace of V of largest possible dimension with* $\langle \, , \, \rangle$ *restricted to* V_+ *positive definite and let* V_- *be a subspace of V of largest possible dimension with* $\langle \, , \, \rangle$ *restricted to* V_- *negative definite. Then* $V = V_+ \oplus V_-$.

Remark B.2.4. The spaces V_+ and V_- are in general not unique.   ◇

Theorem B.2.5 (Sylvester's Law of Inertia). *Let* $\langle \, , \, \rangle$ *be a nonsingular symmetric bilinear form on a real vector space V of dimension t. Let* V_+ *and* V_- *be as in Lemma B.2.3, and let* $r = \dim(V_+)$ *and* $s = \dim(V_-)$. *Then* $\langle \, , \, \rangle$ *is isometric to*

$$r[1] + s[-1].$$

Definition B.2.6. In the situation of Theorem B.2.5, the *signature* $\sigma(\langle \, , \, \rangle) = r - s$.   ◇

Remark B.2.7. Since $r + s = t$, a nonsingular symmetric bilinear form $\langle \, , \, \rangle$ on a real vector space V is determined up to isometry by $\dim(V)$ and $\sigma(\langle \, , \, \rangle)$.   ◇

Lemma B.2.8. *Let* $\langle \, , \, \rangle$ *be a nonsingular symmetric bilinear form on a real vector space V of even dimension t. Suppose that V has a subspace* V_0 *of dimension* $t/2$ *such that the restriction of* $\langle \, , \, \rangle$ *to* V_0 *is identically 0. Then* $\sigma(\langle \, , \, \rangle) = 0$.

Proof. If $v \in V_+ \cap V_0$, and $v \neq 0$, then $\langle v, v \rangle > 0$ as $v \in V_+$, but $\langle v, v \rangle = 0$ as $v \in V_0$, so this is impossible. Hence $V_+ \cap V_0 = \{0\}$, and so $r + t/2 \leq t$. Similarly $V_- \cap V_0 = \{0\}$ so $s + t/2 \leq t$. But $r + s = t$, so we must have $r = s = t/2$, and hence $\sigma(\langle \, , \, \rangle) = 0$.□

Appendix C
Categories and Functors

Category theory provides a very convenient, and for some purposes essential, formulation for algebraic topology. We have minimized its use in this book, but we give the basics here.

C.1 Categories

Definition C.1.1. A *category* $\mathfrak{C}$ consists of a class of *objects* $\mathrm{Obj}(\mathfrak{C}) = \{A, B, C, \ldots\}$ and for any ordered pair (A, B) of objects a class of *morphisms* $\mathrm{Mor}(A, B)$, with the following properties:

(1) Given any $f \in \mathrm{Mor}(A, B)$ and $g \in \mathrm{Mor}(B, C)$ there is their *composition* $gf \in \mathrm{Mor}(A, C)$, and furthermore composition is *associative*, i.e., given $f \in \mathrm{Mor}(A, B)$, $g \in \mathrm{Mor}(B, C)$, and $h \in \mathrm{Mor}(C, D)$, then $h(gf) = (hg)f$.
(2) Given any object A of $\mathfrak{C}$ there is the *identity* morphism $\mathrm{id}_A \in \mathrm{Mor}(A, A)$, and, given any pair of objects A and B, if $f \in \mathrm{Mor}(A, B)$, then $f\mathrm{id}_A = f = \mathrm{id}_B f$. ◊

Example C.1.2.   (1) $\mathrm{Obj}(\mathfrak{C}) = \{\text{sets}\}$ and for $X, Y \in \mathfrak{C}$, $\mathrm{Mor}(X, Y) = \{\text{functions } f : X \to Y\}$.
(2) $\mathrm{Obj}(\mathfrak{C}) = \{\text{topological spaces}\}$ and for $X, Y \in \mathfrak{C}$, $\mathrm{Mor}(X, Y) = \{\text{continuous maps } f : X \to Y\}$.
(3) $\mathrm{Obj}(\mathfrak{C}) = \{(X, A) \mid X \text{ is a topological space and } A \text{ is a subspace of } X\}$ and $\mathrm{Mor}((X, A), (Y, B)) = \{\text{continuous maps } f : X \to Y \text{ with } f(A) \subseteq B\}$.
(4) $\mathrm{Obj}(\mathfrak{C}) = \{\text{abelian groups}\}$ and for $G, H \in \mathfrak{C}$, $\mathrm{Mor}(G, H) = \{\text{group homomorphisms } f : G \to H\}$.
(5) $\mathrm{Obj}(\mathfrak{C}) = \{\text{graded abelian groups } \{G_i\}_{i \in \mathbb{Z}}\}$ and for $\{G_i\}, \{H_i\} \in \mathfrak{C}$, $\mathrm{Mor}(G, H) = \{\{f_i\} \mid f_i : G_i \to H_i \text{ is a group homomorphism}\}$.

© Springer International Publishing Switzerland 2014
S.H. Weintraub, *Fundamentals of Algebraic Topology*, Graduate Texts
in Mathematics 270, DOI 10.1007/978-1-4939-1844-7

(6) $\text{Obj}(\mathfrak{C}) = \{$chain complexes $\{G_i, \partial_i^G : G_i \to G_{i-1}\}\}$ and $\text{Mor}(\{G_i, \partial_i^G\}, \{H_i, \partial_i^H\}) = \{\{f_i\} \mid f_i : G_i \to H_i$ is a group homomorphism with $\partial_{i-1}^H f_i = f_{i-1} \partial_i^G\}$.

(7) $\text{Obj}(\mathfrak{C}) = \{$graded rings $\{G^i\}\}$. A graded ring has the additive structure of a graded abelian group and a multiplicative structure $u_G : G^i \otimes G^j \to G^{i+j}$ satisfying:

  (a) There is an identity element $1^G \in G^0$.
  (b) Multiplication is *commutative* in the sense that if $\alpha \in G^i$ and $\beta \in G^j$, then $\alpha\beta = (-1)^{ij}\beta\alpha$.

  $\text{Mor}(\{G^i\}, \{H^i\}) = \{\{f^i\}\}$ where $\{f^i\}$ is a homomorphism of graded rings, i.e., a morphism of graded abelian groups that is also a homomorphism of rings with identity.

(8) $\text{Obj}(\mathfrak{C}) = \{$CW-complexes$\}$ and $\text{Mor}(X,Y) = \{$cellular maps $f : X \to Y\}$, where $f : X \to Y$ is cellular if $f(X^n) \subseteq Y^n$ for every n.

(9) $\text{Obj}(\mathfrak{C}) = \{$pointed spaces$\} = \{(X, x_0)\}$, i.e., a nonempty topological space X and a point $x_0 \in X$, and $\text{Mor}((X, x_0), (Y, y_0)) = \{f : X \to Y \mid f(x_0) = y_0\}$.

(10) $\text{Obj}(\mathfrak{C}) = \{$groups$\}$ and $\text{Mor}(G, H) = \{$group homomorphisms $f : G \to H\}$.

(11) $\text{Obj}(\mathfrak{C}) = \{$cochain complexes $\{G^i, \delta_G^i : G^i \to G^{i+1}\}\}$ and $\text{Mor}(\{G^i, \delta_G^i\}, \{H^i, \delta_H^i\}) = \{\{f^i\} \mid f^i : G^i \to H^i$ is a group homomorphism with $\delta_H^{i+1} f_i = f_{i+1} \delta_G^i\}$. $\qquad\qquad\qquad\qquad\qquad\qquad\qquad\qquad\qquad\qquad\qquad\qquad \diamondsuit$

C.2   Functors

Given two categories $\mathfrak{C}$ and $\mathfrak{D}$, we may regard them each as objects and ask for the appropriate notion of a function between them. This notion is that of a functor, and comes in two varieties.

Definition C.2.1. Let $\mathfrak{C}$ and $\mathfrak{D}$ be categories. A *covariant functor* $T : \mathfrak{C} \to \mathfrak{D}$ consists of:

(1) A function $T : \text{Obj}(\mathfrak{C}) \to \text{Obj}(\mathfrak{D})$.

(2) A function $T : \text{Mor}(C_1, C_2) \to \text{Mor}(D_1, D_2)$, where C_1 and C_2 are objects of $\mathfrak{C}$, and $D_1 = T(C_1)$, $D_2 = T(C_2)$, objects of $\mathfrak{D}$, with the properties:

  (a) $T(\text{id}_C) = \text{id}_{T(C)}$ for any object C of $\mathfrak{C}$.
  (b) If C_1, C_2, and C_3 are objects of $\mathfrak{C}$, $f \in \text{Mor}(C_1, C_2)$, and $g \in \text{Mor}(C_2, C_3)$, then $T(gf) = T(g)T(f)$.

A *contravariant functor* $T : \mathfrak{C} \to \mathfrak{D}$ consists of:

(1) A function $T : \text{Obj}(\mathfrak{C}) \to \text{Obj}(\mathfrak{D})$.

(2) A function $T : \text{Mor}(C_1, C_2) \to \text{Mor}(D_2, D_1)$, where C_1 and C_2 are objects of $\mathfrak{C}$, and $D_1 = T(C_1)$, $D_2 = T(C_2)$, objects of $\mathfrak{D}$, with the properties:

(a) $T(\mathrm{id}_C) = \mathrm{id}_{T(C)}$ for any object C of $\mathfrak{C}$.
(b) If C_1, C_2, and C_3 are objects of $\mathfrak{C}$, $f \in \mathrm{Mor}(C_1, C_2)$, and $g \in \mathrm{Mor}(C_2, C_3)$, then $T(gf) = T(f)T(g)$. $\Diamond$

In the following example, we just give the objects of the categories involved. The morphisms should be clear.

Example C.2.2. (1) Let $\mathfrak{C} = \{$pointed spaces$\}$ and $\mathfrak{D} = \{$groups$\}$. We have the covariant functor $T : \mathfrak{C} \to \mathfrak{D}$ given by $T(X, x_0) = \pi_1(X, x_0)$. If $f : (X, x_0) \to (Y, y_0)$ then $T(f) = f_* : \pi_1(X, x_0) \to \pi_1(Y, y_0)$.

(2) Let $\mathfrak{C} = \{(X, A)\}$ pairs of topological spaces and let $\mathfrak{D} = \{$graded abelian groups$\}$. We have the covariant functor $T : \mathfrak{C} \to \mathfrak{D}$ given by $T(X, A) = \{H_i(X, A)\}$, and if $f : (X, A) \to (Y, B)$, $T(f) = \{f_i : H_i(X, A) \to H_i(Y, B)\}$, for any fixed homology theory. Similarly, we have the contravariant functor $T : \mathfrak{C} \to \mathfrak{D}$ given by $T(X, A) = \{H^i(X, A)\}$, and if $f : (X, A) \to (Y, B)$, $T(f) = \{f^i : H^i(Y, B) \to H^i(X, A)\}$, for any fixed cohomology theory.

(3) In this and the remaining examples, we simply describe what T does on objects; its effect on morphisms should then be clear.

Let $\mathfrak{C} = \{$chain complexes$\}$ and $\mathfrak{D} = \{$graded abelian groups$\}$. Then $T : \mathfrak{C} \to \mathfrak{D}$ by $T(\{C_i\}) = \{H_i\}$ where $\{H_i\}$ are the homology groups of $\{C_i\}$. Similarly, if $\mathfrak{C} = \{$cochain complexes$\}$ we have $T : \mathfrak{C} \to \mathfrak{D}$ by $T(\{C^i\}) = \{H^i\}$ where $\{H^i\}$ are the cohomology groups of $\{C^i\}$. Note in both cases T is covariant.

(4) Let $\mathfrak{C} = \{$chain complexes$\}$ and $\mathfrak{D} = \{$cochain complexes$\}$. Let $T : \mathfrak{C} \to \mathfrak{D}$ be the contravariant functor given by $T(\{C_i\}) = $ the dual cochain complex $\{C^i = \mathrm{Hom}(C_i, \mathbb{Z})\}$.

(5) Let $\mathfrak{C} = \{$topological pairs $(X, A)\}$ and let $\mathfrak{D} = \{$chain complexes$\}$. Let $T : \mathfrak{C} \to \mathfrak{D}$ be the covariant functor given by $T(X, A) = \{C_i(X, A)\}$, the singular chain complex of the pair (X, A).

(6) Let $\mathfrak{C} = \{(X, A)\}$ pairs of topological spaces and let $\mathfrak{D} = \{$graded rings$\}$. We have the contravariant functor $T : \mathfrak{C} \to \mathfrak{D}$ given by $T(X, A) = \{H^i(X, A)\}$ where $H^i(X, A)$ denotes singular cohomology. (We have constructed a ring structure on singular cohomology in Sect. 5.6.)

(7) Let $\mathfrak{C} = \{$CW-complexes$\}$ and $\mathfrak{D} = \{$graded abelian groups$\}$. Then $T : \mathfrak{C} \to \mathfrak{D}$ by $T(X) = \{H_i^{\mathrm{cell}}(X)\}$ is a covariant functor and $T : \mathfrak{C} \to \mathfrak{D}$ by $T(X) = \{H^i_{\mathrm{cell}}(X)\}$ is a contravariant functor. Note that $f \in \mathrm{Mor}(X, Y)$ induces maps on cellular homology or cohomology as by the definition of $\mathfrak{C}$, $\mathrm{Mor}(X, Y)$ consists of cellular maps. $\Diamond$

Remark C.2.3. Suppose that $H_i(X, A)$ is singular homology, and that $H^i(X, A)$ is singular cohomology. Then the covariant functor in Example C.2.2(2) is in an obvious way the composition of the covariant functor in (5) with the covariant functor in (3), and the contravariant functor in (2) is in an obvious way the composition of the covariant functor in (5), the contravariant functor in (4), and the covariant functor in (3). Note that the functors in (3) and (4) are purely algebraic. It is the functor in (5) that makes the connection between topology and algebra. $\Diamond$

Bibliography

Here is a short annotated guide to various other books on algebraic topology.

1. J. F. Davis and P. Kirk, *Lecture Notes in Algebraic Topology*, American Mathematical Society, 2001.
   Roughly speaking, this book picks up more or less where our book leaves off.
2. S. Eilenberg and N. Steenrod, *Foundations of Algebraic Topology*, Princeton University Press, 1952.
   This monograph introduced the Eilenberg-Steenrod axioms, which codified homology and cohomology theory and laid the foundation for future developments. It is very clearly written and has stood the test of time well.
3. B. Gray, *Homotopy Theory: An Introduction to Algebraic Topology*, Academic Press, 1975.
   As its title indicates, an exposition of algebraic topology from the viewpoint of homotopy (rather than homology) theory.
4. M. J. Greenberg and J. R. Harper, *Algebraic Topology: A First Course*, Westview Press, 1981.
   The first edition of this book (written by Greenberg alone and published in 1967) was this author's introduction to the subject.
5. A. Hatcher, *Algebraic Topology*, Cambridge University Press, 2002.
   A wide-ranging modern text, written from a geometric viewpoint. It can also be downloaded, free and legally, from the author's website.
6. S. MacLane, *Homology*, Springer, 1963.
   A clear and thorough introduction to homological algebra, written by one of the giants in the field.
7. W. S. Massey, *A Basic Course in Algebraic Topology*. Springer, 1991.
   An excellent introductory text written by a leading algebraic topologist. While Massey's book does not adopt an axiomatic standpoint, its spirit is in many ways similar to ours.
8. E. H. Spanier, *Algebraic Topology* McGraw-Hill, 1966.
   This book was and remains the bible of the subject, as it existed at the time of its writing. Precisely because it is so thorough, and written in the greatest generality, it is not easy to read, but is an invaluable reference.

© Springer International Publishing Switzerland 2014
S.H. Weintraub, *Fundamentals of Algebraic Topology*, Graduate Texts
in Mathematics 270, DOI 10.1007/978-1-4939-1844-7

Index

© Springer International Publishing Switzerland 2014
S.H. Weintraub, *Fundamentals of Algebraic Topology*, Graduate Texts
in Mathematics 270, DOI 10.1007/978-1-4939-1844-7

Printed in the United States
By Bookmasters